SAGE was founded in 1965 by Sara Miller McCune to support the dissemination of usable knowledge by publishing innovative and high-quality research and teaching content. Today, we publish more than 750 journals, including those of more than 300 learned societies, more than 800 new books per year, and a growing range of library products including archives, data, case studies, reports, conference highlights, and video. SAGE remains majority-owned by our founder, and after Sara's lifetime will become owned by a charitable trust that secures our continued independence.

Los Angeles | London | Washington DC | New Delhi | Singapore | Boston

Achieving Universal Energy Access in India

Achieving Universal Energy Access in India
Challenges and the Way Forward

P.C. Maithani
Deepak Gupta

www.sagepublications.com
Los Angeles • London • New Delhi • Singapore • Washington DC • Boston

Copyright © P.C. Maithani and Deepak Gupta, 2015

All rights reserved. No part of this book may be reproduced or utilised in any form or by any means, electronic or mechanical, including photocopying, recording, or by any information storage or retrieval system, without permission in writing from the publisher. The views expressed in the book are the personal views of the authors and in no way should be construed as that of the government.

First published in 2015 by

SAGE Publications India Pvt Ltd
B1/I-1 Mohan Cooperative Industrial Area
Mathura Road, New Delhi 110 044, India
www.sagepub.in

SAGE Publications Inc
2455 Teller Road
Thousand Oaks, California 91320, USA

SAGE Publications Ltd
1 Oliver's Yard, 55 City Road
London EC1Y 1SP, United Kingdom

SAGE Publications Asia-Pacific Pte Ltd
3 Church Street
#10-04 Samsung Hub
Singapore 049483

Published by Vivek Mehra for SAGE Publications India Pvt. Ltd, typeset in 10/13pt Berkeley by Diligent Typesetter, Delhi and printed at Sai Print-o-Pack, New Delhi.

Library of Congress Cataloging-in-Publication Data

Maithani, P. C.
   Achieving universal energy access in India : challenges and the way forward / P. C. Maithani, Deepak Gupta.
     pages cm
   Includes bibliographical references and index.
   1. Rural electrification—India. 2. Energy development—India. 3. Renewable energy resources—India. 4. Energy policy—India. 5. Rural development—India. I. Title.
   HD9688.I52M35   333.793'2—dc23   2015   2014049341

ISBN: 978-93-515-0137-4 (HB)

The SAGE Team: Rudra Narayan, Alekha Chandra Jena and Anju Saxena

*To
Millions of rural households in India who
are in need of energy access*

Thank you for choosing a SAGE product! If you have any comment, observation or feedback, I would like to personally hear from you. Please write to me at contactceo@sagepub.in

—Vivek Mehra, Managing Director and CEO,
SAGE Publications India Pvt Ltd, New Delhi

Bulk Sales

SAGE India offers special discounts for purchase of books in bulk. We also make available special imprints and excerpts from our books on demand.

For orders and enquiries, write to us at

Marketing Department
SAGE Publications India Pvt Ltd
B1/I-1, Mohan Cooperative Industrial Area
Mathura Road, Post Bag 7
New Delhi 110044, India
E-mail us at marketing@sagepub.in

Get to know more about SAGE, be invited to SAGE events, get on our mailing list. Write today to marketing@sagepub.in

This book is also available as an e-book.

Contents

List of Tables	xi
List of Figures	xiii
List of Abbreviations	xv
Preface	xxi

1	The Importance of Energy Access	1
	Background	1
	Human Development and Energy Access	4
2	Global Status	10
	Global Scenario	10
	International Efforts for Energy Access	12
3	India Electricity Status	20
	Background	20
	Current Electricity Mix	22
	Future Projections	23
	Cost of Power	33
	Conclusion	40
4	Rural Electrification: Policy Landscape and Status of Access	43
	Policies (1974–2005)	45
	Electricity Act 2003	48
	Rajiv Gandhi Grameen Vidyutikaran Yojana	49
	Electricity Access Programmes and RGGVY: An Assessment	56
	Issues in Rural Electrification	60

5 Renewable Energy Options for Electricity Access — 73
- The Remote Village Electrification Programme — 74
- Bank-financed Solar Lighting Programme — 81
- Mini-grid Model — 90
- Lanterns — 108
- Franchisee Model — 110
- Localised Grid-based Supplies — 111
- Some Examples and Models — 112
- Cost Competitiveness of Off-grid Renewable Energy Systems — 115
- Technology — 116

6 Challenges for Universal Electricity Access and Way Forward — 121
- Background — 121
- Sociological or Organisational Challenges — 124
- Entrepreneurial Challenges — 125
- Technical Challenges — 128
- Regulatory Barriers — 131
- Issues Related to Tariff, Costs and Subsidy Support — 131
- Grid Interactivity — 137
- Access to Capital — 139
- Bilateral/Multilateral Agencies — 143
- An Alternative Framework to Achieve Universal Electricity Access — 144
- Actions Required — 148

7 Access to Cooking Energy — 163
- The Problem — 163
- Health Impacts — 164
- Other Effects — 169
- IAP and Poverty — 171
- Policy Implications — 172
- Extent of Global Burden — 174
- Status in India — 175
- Resource Availability of Fuels — 179
- National Efforts — 180
- Programmes in India — 183

Contents ix

National Project on Biogas Development	188
Focus on LPG	193
Solar Cooking	195
Lessons Learned	196
New National Programme	199

8 Subsidies and Funding — 205
- Cooking Energy: Inter-fuel Substitution in Rural Areas — 205
- Difficulties of LPG — 206
- Kerosene — 207
- Funding and Subsidies — 212
- Subsidy for Biogas — 213
- Financing of Subsidy — 215
- The Carbon Market — 216
- International Funding — 218
- Domestic Funding — 219
- The Kerosene Conundrum — 219

9 Energy Access and Rural Development — 229
- Rural Energy Market — 229
- Agricultural Pump Sets — 236
- Impact — 239

10 The Last Word — 241

Bibliography — 249
Index — 256
About the Authors — 262

List of Tables

2.1	Number of People without Access to Electricity and Relying on Traditional Use of Biomass (in Millions)	12
2.2	Generation Requirement for Universal Electricity Access, 2030 (TWh)	17
3.1	Viability of Major State Utilities (Excluding Delhi and Odisha)	37
4.1	Status on RGGVY Progress during 10th and 11th Plans	53
4.2	Village Electrification Status as on 31 January 2014	53
4.3	State-wise Rural Household Electrification Levels	56
5.1	Year-wise State-wise Remote Village Electrification	76
5.2	State/Bank-wise Data of Solar Lighting Units (as on 31 January 2014)	83
5.3	DDG Projects under RGGVY (as on February 2014)	94
5.4	State-wise Cumulative Solar Lanterns (as on 31 December 2013)	108
7.1	Health Effects of Biomass Fuel Use in Cooking	166
7.2	State-wise Distribution of Households by Type of Fuel Used for Cooking (Per Cent)	177
7.3	Standard Performance Parameters for Cookstoves	187
7.4	Family Type Biogas Plants: State-wise Estimated Potential and Cumulative	189
7.5	Year-wise Achievement for Family Type Biogas Plants during the 11th Five Year Plan	191
7.6	Approved Models of Cookstoves	201

8.1 State-wise Kerosene Allocation (in MT) 221
8.2 Under-recoveries of OMCs and Compensation by
    Upstream Companies and the Government 224
8.3 Actual Retail Selling Price of PDS Kerosene and
    Under-recovery at Delhi 226

List of Figures

1.1	Relationship of HDI and Per Capita Electricity Consumption	5
1.2	The Relationship between Per Capita Final Energy Consumption and Income in Developing Countries	7
1.3	Comparison of the EDI to the HDI	7
1.4	Per Capita Energy Consumption and Emission in Selected Regions	8
2.1	Share of Population (Per Cent) without Access to Electricity	11
2.2	Share of Population (Per Cent) Relying on Traditional Biomass for Cooking	11
3.1	India Has Low Per Capita Energy, Electricity Consumption and CO_2 Emission	21
3.2	Percentage Share of Renewable Power in the Electricity Installed Capacity	22
3.3	Percentage Share of Renewable Energy in Electricity Mix	23
3.4	Renewable Power Installed Capacity	24
3.5	Technology-wise Electricity Mix 2013–14	24
3.6	Discoms—Trends of Subsidy Dependence	38
4.1	Households by Main Source of Lighting	63
5.1	VEC Model	92
6.1	MNRE Subsidy for Bank-financed Projects—No. of Units	153
7.1	Incidences of Respiratory Symptoms for Males and Females by Age Group	167

7.2 Premature Annual Deaths from Household Air Pollution
and Other Diseases — 168
7.3 Households' Main Source of Cooking — 176

8.1 (a) Total Rural Energy Consumption Showing Source Share, by Income Decile and (b) End-use Rural Energy Consumption Showing Source Share, by Income Decile — 209

List of Abbreviations

ACS	average cost of supply
ADATS	Agriculture Development and Training Society
ADB	Asian Development Bank
AEC	Atomic Energy Commission
AGECC	Advisory Group on Energy and Climate Change
AHBPPL	Arashi Hi-Tech Bio-Power Private Limited
ALRI	acute lower respiratory infection
ANMs	auxiliary nurse midwives
APL	above poverty line
APP	Association of Power Producers
AQGs	Air Quality Guidelines
AREP	Accelerated Rural Electrification Programme
ARI	acute respiratory infection
ARR	average revenue realised
ARTI	Appropriate Rural Technology Institute
ATM	automatic teller machine
bcm	billion cubic meters
BIS	Bureau of Indian Standards
BOS	balance of system
BP	British Petroleum
BPL	below poverty line
BREDA	Bihar Renewable Energy Development Agency
BTS	base transceiver station
CDM	clean development mechanism
CEA	Central Electricity Authority
CEL	Central Electronics Limited
CERC	Central Electricity Regulatory Commission
CFA	central financial assistance

CFL	compact fluorescent lamp
CGTMSE	Credit Guarantee Trust for Medium and Small Enterprises
CHC	community health centre
CIL	Coal India Limited
CONCOR	Container Corporation of India
COPD	chronic obstructive pulmonary disease
CPSE	Central Public Sector Enterprise
CREDA	Chhattisgarh Renewable Energy Development Agency
CRISIL	Credit Rating Information Services of India Limited
CSE	Centre for Science and Environment
CSIR	Council of Scientific and Industrial Research
CSR	Corporate Social Responsibility
DAE	Department of Atomic Energy
DALY	Disability-adjusted Life Year
DDG	Decentralized Distributed Generation
DG	diesel generator
Discoms	distribution companies
DPR	detailed project report
EDI	Energy Development Index
EIU	Economist Intelligence Unit
ESMAP	Energy Sector Management Assistance Programme
EU	European Union
FCCC	Framework Convention on Climate Change
FY	financial year
GAM	Grameena Abhivrudhi Mandali
GBI	generation-based incentive
GDP	gross domestic product
GHG	greenhouse gas
GIZ	Deutsche Gesellschaft für Internationale Zusammenarbeit
GTZ	Deutsche Gesellschaft für Technische Zusammenarbeit
GW	giga watt
HDI	Human Development Index
HLS	home lighting system
HPS	Husk Power Systems
IAP	indoor air pollution
ICMR	Indian Council of Medical Research, New Delhi
ICRA	Indian Credit Ratings Agency

List of Abbreviations xvii

IEP	Integrated Energy Policy
IEX	Indian Energy Exchange
IIM	Indian Institute of Management
IIT	Indian Institute of Technology
IMF	International Monetary Fund
IMMT	Institute of Minerals & Materials Technology
IOC	Indian Oil Corporation
IOREC	International Off-grid Renewable Energy Conference
IRADe	Integrated Research and Action for Development
IREDA	Indian Renewable Energy Development Agency
IRENA	International Renewable Energy Agency
IRR	internal rate of return
ITC	India Tobacco Company
ITI	Industrial Training Institutes
IUATLD	International Union Against Tuberculosis and Lung Disease
J&K	Jammu & Kashmir
JNNSM	Jawaharlal Nehru National Solar Mission
KERC	Kartnataka Electricity Regulatory Commission
KPMG	Klynveld Peat Marwick Goerdeler
KPTCL	Karnataka Power Transmission Corporation Limited
kV	kilovolt
L&T	Larsen & Toubro
LaBL	Lighting a Billion Lives
LED	light emitting diode
LNG	liquefied natural gas
LPG	liquid petroleum gas
LWE	Left Wing Extremism
MDGs	Millennium Development Goals
MNP	Minimum Needs Program
MNRE	Ministry of New and Renewable Energy
MGNREGA	Mahatma Gandhi National Rural Employment Guarantee Act
MOP	Ministry of Power
MT	metric tonne
NABARD	National Agricultural Bank and Rural Development
NAPCC	National Action Plan on Climate Change

NALCO	National Aluminium Company Limited
NBMMP	National Biogas and Manure Management Programme
NCAER	National Council of Applied Economic Research
NCD	non-communicable disease
NCI	National Improved Cookstove Initiative
NPBD	National Project on Biogas Development
NE	North East
NEP	National Electricity Policy
NGO	non-governmental organisation
NPA	non-performing assets
NPBD	National Project on Biogas Development
NPIC	National Programme on Improved Cookstoves
NRECA	National Rural Electric Co-operatives Association
NRI	non-resident Indian
NSS	National Sample Survey
NSSO	National Sample Survey Organisation
NTPC	National Thermal Power Corporation
O&M	operations and maintenance
OMC	Omnigrid Micropower Company
OMC	Oil Marketing Company
ONGC	Oil and Natural Gas Corporation Ltd
OPEC	Organization of Petroleum Exporting Countries
ORF	Observer Research Foundation
PDS	Public Distribution System
PEACE	Promoting Energy Access through Clean Energy
PFC	Power Finance Corporation
PHC	primary health centres
PhD	Doctor of Philosophy
PLB	poverty link basket
PMGY	Pradhan Mantri Gramoday Yojana
PSU	public sector undertaking
R&D	research and development
RBI	Reserve Bank of India
REC	Rural Electrification Corporation
REDB	Rural Electricity Distribution Backbone
REDF	Renewable Energy Development Fund
REST	Rural Electricity Supply Technology

List of Abbreviations xix

RET	Renewable Energy Technology
RGGLVY	Rajiv Gandhi Gramin LPG VitaranYojana
RGGVY	Rajiv Gandhi Grameen Vidyutikaran Yojana
RIDF	Rural Infrastructure Development Fund
RPCD	Rural Planning and Credit Department
RSP	retail selling price
RVE	Remote Village Electrification
SC	scheduled caste
SE4ALL	Sustainable Energy for All
SEB	State Electricity Board
SECI	Solar Energy Corporation of India
SELCO	Solar Electric Light Company
SERC	State Electricity Regulatory Commission
SEWA	Self Employed Women's Association
SHG	self-help group
SHS	solar home system
SIDBI	Small Industries Development Bank of India
SPPS	single point power supply
SPV	solar photovoltaic
SRE	Saran Renewable Energy
SREDCOP	Sagar Rural Energy Development Cooperative
ST	scheduled tribe
TB	tuberculosis
TBU	technical backup unit
TCES	total commercial energy supply
TERI	The Energy and Resources Institute
TIDE	Technology Informatics Design Endeavour
TNEB	Tamil Nadu State Electricity Board
toe	tonne of oil equivalent
TV	television
TWh	terra watt hour
UHNI	ultra-high net worth individual
UK	United Kingdom
UN	United Nations
UNCSD	United Nations Conference on Sustainable Development
UNDP	United Nations Development Programme
UNIDO	United Nations Industrial Development Organisation

UPA	United Progressive Alliance
US AID	United States Agency for International Development
USA	United States of America
UTs	Union Territories
VEC	Village Energy Committees
VEI	Village Electrification Infrastructure
VESP	Village Energy Security Programme
WBREDA	West Bengal Renewable Energy Development Agency
WEO	World Energy Outlook
WHO	World Health Organization
Wp	watt-peak
WSSD	World Summit on Sustainable Development
WTP	willingness to pay

Preface

The last decade of the last century saw some unshackling of the barriers to growth in India. The first decade of this century saw the rising hopes and aspirations of Indians as we went through a period of rapid growth. There was great optimism and reasonable confidence that we would have inclusive development and substantive alleviation of poverty. Although the last few years have witnessed declining growth for various reasons, there is little doubt that India will be on a high growth trajectory.

Although we have made good progress over these years, and also made a noticeable impact on poverty, there is an impression that inequality has increased. We also seem to have embarked upon an energy-intensive consumerist model of development which tends to cater to the demands and lifestyles of the rich and the upper middle class, even as millions of poor Indians, primarily in rural India, continue to live lives of distress and without the basic necessities of life.

While this dichotomy can be generally seen in many sectors, this is particularly acute in the area of energy. India has amongst the lowest energy consumptions per capita. This is because energy access is so poor and energy supply so inadequate even as the consumption of developed India is not much less than the consumption of China or the developed world.

In the earlier years, the power sector was somewhat neglected. However, since the beginning of this century, the power sector got a lot of attention starting with the passing of the Electricity Act in 2003. Capacity creation also substantially increased in the last decade. The RGGVY sought to address the problem of rural electrification and substantial progress was made. Nevertheless we have not been able to set up enough electricity generation capacity, supply and distribute efficiently what we generate and charge enough to ensure viability. Consequently, we continue to have power cuts in urban areas, both big and small, even as parts of rural India are virtually in the dark, even though over 95 per cent of the villages have been electrified. We have not been able to sort out the real problems of rural electrification in such areas. The situation is also unlikely to improve soon, or perhaps even in the longer term.

The problem is equally acute with regard to cooking energy. We are committed to supply modern fuels but progress has been poor even as the urban area and the rural rich enjoy unsustainable subsidies. This problem is also unlikely to go away.

Over the years it has been presumed that electricity from the grid and cooking gas would be possible to be supplied to every household. The central theme of the book is to discuss the hard reality that this is unlikely to happen.

We, therefore, discuss and examine the alternatives of supplying electricity through renewable energy sources, which have now become possible, and the necessity of having improved biomass cookstoves for cooking. We examine what has happened, analyse the constraints and opportunities and suggest what is possible as well as the measures which need to be taken, difficult and complex as they may be.

There has been little discussion of energy access issues in policy-making institutions. There has, therefore, not been much analysis of why we have failed, and even the extent of the failure. There has also been little public debate. Whatever little there has been, it is confined to seminars here and there, and small pilot projects, also here and there. Our attempt in this book is both to contribute to a constructive policy and public debate which we hope will take place and to underline the urgency for it to happen. This is especially so in the context of energy access becoming one of the sustainable development goals to be adopted by the UN in 2015.

We have also tried to place the energy access issue in the larger context of rural development, particularly for our most backward areas, as well as the moral imperatives of equitable and inclusive growth. This gives a larger dimension to this issue as well as suggests that it be given the highest policy priority.

Currently there is an attempt to re-invigorate the power sector, accelerate development of renewable energy resources, increase coal production, target 24×7 supply of electricity to all households by 2019, and to revisit subsidies and the allocation of kerosene. The recent decline in global oil prices has, of course, reduced subsidy burdens on kerosene. The recently launched Deendayal Upadhyaya Gram Jyoti Yojana (DDUGJY) plans to separate agricultural and non-agricultural feeders for judicious rostering of supply to agricultural and non-agricultural consumers in rural areas and strengthening and augmentation of sub-transmission and distribution infrastructure in rural areas, including metering of distribution transformers/feeders/consumers. These measures should lead to improvement in hours of power supply in rural areas. In this context also, we hope this book would help provide sharper focus to the issue of energy access.

This is primarily an academic study, but written with the hope that it may help policy formulation. We have tried to extensively explore the literature on the subject. A lot has been written and discussed at different places and in parts. We have tried to present a holistic picture and covered both electricity and cooking energy access. This should help both policy and academic discussions as well as prove useful to researchers.

The Ministry of New and Renewable Energy in the last few years gave us the opportunity and experience to view these problems and issues with a different perspective. It has also given the confidence that these problems are resolvable. This book is our attempt to put these in the public domain.

P.C. Maithani
Deepak Gupta

1
The Importance of Energy Access

Background

Energy has been the fundamental driver of the modern economy. The discovery of fossil fuels, and the development of technology and its various applications to exploit their use, led to the widespread availability of affordable energy, which laid the foundation of the industrial revolution in the developed countries. The developing countries followed the process much later in the day and are now trying to catch up with them, but the times and circumstances are not the same, with many serious new and emerging challenges and constraints. Today, energy security and dealing with energy's adverse impact on climate change, individually for most nations, and globally as a comity of nations, are considered to be the two over-riding challenges faced by the energy sector on the road to a sustainable future. Energy is thus increasingly becoming central to the whole debate on sustainable development, being critical to the current economic, environmental and developmental issues facing the world. Much of the global discussion is focussed on these two issues. Developing countries face two additional challenges. First, how do they generate increased, reliable and affordable energy services for their growth and prosperity? This involves the traditional questions of financing the generation and transmission infrastructure needed to increase higher electricity penetration while simultaneously ensuring adequate cost recovery for long-term viability. Much of our national attention is concentrated on this, as indeed it should, since this is a major problem of our power sector. Second, how do many of these countries simultaneously resolve the problem of energy access, which will become fundamental to the future of their inclusive growth and critical for both reduction of poverty and improvement in the health and quality of life of a large portion of their citizens living in rural areas? This also brings forth the equity dimension of energy. This is an area, however, of comparative neglect but one that requires our urgent attention.

Why is energy access so important and what does it mean? While there is no internationally adopted definition of energy access as yet, it could broadly be defined as the physical availability of modern energy carriers and improved end-use devices at the household level at affordable prices. It includes access to less polluting household energy for cooking and heating (improved cookstoves with traditional solid biomass fuels, liquid and gaseous fuels such as kerosene and liquid petroleum gas [LPG]), or electricity for powering appliances and lights in households and public facilities, which could come from renewable sources, and mechanical power from either electricity or other energy sources that improve the productivity of labour (Pachauri et al., 2012). The Report of the UN Advisory Group on Energy and Climate Change describes it as 'access to clean, reliable and affordable energy services for cooking and heating, lighting, communication and productive uses' (United Nations, 2010). It has elaborated this by stating that the levels of energy access should be such that it can improve livelihoods in the poorest countries and drive local economic development. Improvement really means lighting for households and help in communication, education, health and perhaps some entertainment, and cleaner sources of cooking. Productive uses mean water pumping, cottage industry and agricultural processing, which would both improve livelihoods and drive economic development. 'Affordable' in this context has been explained so that the 'cost to the end users is compatible with their income levels and no higher than the cost of traditional fuels, in other words what they would be able to and willing to pay for the increased quality of energy supply'.[1] Reliability, adequacy or quantity, timeliness and quality are also important parts—provision of infrastructure or actual supply is not enough unless power is delivered when it is needed and can be properly used.²

Thus, energy access has to be discussed in a larger context. Energy is a cross-cutting overarching input that makes possible effective delivery of public services thus having direct impact on practically all aspects of human welfare and human development, including access to water, agricultural productivity, healthcare, education and communication. Access to affordable and reliable energy services, therefore becomes fundamental to better standards of household living, reducing poverty and improving health, increasing productivity, enhancing competitiveness and promoting economic growth (IEA, 2011).

Lighting in low income households in developing countries where rural grid power supply is poor is generally provided by kerosene in low-efficiency lanterns. Kerosene is unsafe, produces a dim light, making it hard to do school work or house work at night, and it produces noxious fumes. Solar is clearly better, providing both electricity and much brighter light through compact fluorescent lamps (CFLs) and light emitting diodes (LEDs) by operating a switch and having no harmful emissions. Thus, electricity access through solar provides immediate beneficial impact in the lighting and general environment inside the house, a great value in itself. It allows children to study relatively comfortably for longer hours, a huge benefit reported by practically all households in informal conversations. Benefits include easier atmosphere for cooking/household chores and to do some income-generating activities for longer hours by women. Cooking on traditional stoves makes women and infants suffer from a number of respiratory diseases because of their exposure to large amounts of smoke and particulates. It also makes women spend a significant amount of time and effort to collect biomass fuel. Access to improved stoves would provide substantial relief.

It is important to stress that one may not consider these benefits to be very significant when it is seen from the point of view of an ordinary urban household having several bulbs for lighting and LPG for cooking, because we are unable to transcend our immediate environment and imagine a life without these comforts, which are taken for granted. But for the poor rural household, the quality of life improves tremendously. It is critical that we (and the policy makers) put a high value to this.³

Human Development and Energy Access

In the above context, it would be useful to consider the relationship between human development and energy access. The Human Development Index (HDI) is now widely considered as a better measure of development than gross domestic product (GDP). It is composed of data on life expectancy, education, per capita GDP and other standard-of-living indicators at the national level. It not only includes growth but also the concept of inclusive development, and embedded within, some measure of equity. Typically, human development has been interpreted as 'the expansion of people's freedoms and capabilities to lead lives that they value and have reason to value', which is a more expansive notion than basic needs. The development of human capabilities must be the first priority for three reasons. First, these capabilities are actually ends in themselves. Second, they are also important instrumentalities. Third, it would also ensure our growth is more inclusive. Any discussion on development, therefore, has not only to consider how to meet these needs but also aim to go beyond them to improve capabilities. The noted economist Prof. Amartya Sen, who developed this capability framework, has theorised energy carriers as commodities or input factors that expand an individual's set of capabilities as it provides lighting, motive power and access to mass media and telecommunications. Access to energy services must be seen in this context.

It is also important to understand the relationship between energy and poverty. At the outset, the relationship between household energy consumption and poverty is clearly bi-directional. On the one hand, access to modern energy can contribute to poverty alleviation and improve indicators of well-being such as income, education or access to clean water. On the other hand, lack of access is not only in itself a sign of poverty, but also contributes to it. It may also help to understand what is meant by energy poverty. A World Bank policy research working paper has discussed the issue of energy poverty line. It notes several attempts to define energy poverty such as quantification of a household's direct energy needs; level of energy used by households below the known expenditure or income poverty line; and energy expenditure as percentage of total with 10 per cent being assumed somewhat arbitrarily. It proposes a demand-based approach to define the energy poverty line as the threshold point at which energy consumption begins to increase with increases in income (Khander, 2010).[4] For the poorest, energy use does not increase with rise in incomes because their energy expenditures are already high though they are at the barest minimum. This means lower levels of household welfare because lesser amounts are available for other goods and services. It also means that energy poverty is worse than expenditure poverty in rural areas and energy inequality is greater than income inequality. It also follows that the decrease in energy poverty would be much higher for the poorer households compared to the wealthier ones. A short study examining the data provided by the India Human Development Survey (IHDS) concludes that impact of energy accessibility on educational and health attainment may be even greater than income alleviation. Another study concludes that HDI increases rapidly with slight increase in energy consumption, and this increase gradually lessens with higher consumption levels (Reddy, 2009). This would be evident from Figure 1.1. This suggests the huge gains that can be made in HDI by improving energy access. Provision of energy access thus becomes a critical instrument in improving the condition of the poorest.

This discussion can be extended further. Expenditure-based poverty measures view deprivation through a single lens of income/consumption. But there is a general consensus that this approach has proved to be

Figure 1.1
Relationship of HDI and per capita electricity consumption

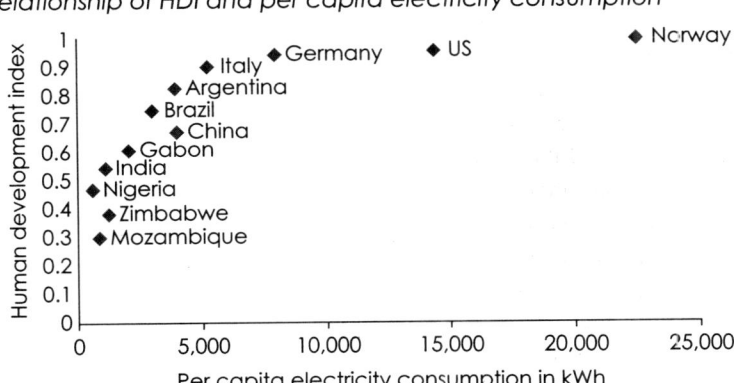

Source: Human Development Report 2011 and IEA Key Energy Statistics 2011.

inadequate in capturing the multiple deprivations that are part and parcel of a life lived in poverty. Where there is deprivation, there are barriers to the advancement of human development (Gupta, 2014). This has led to the development of a multi-dimensional Poverty Index—a measure of poverty that captures the hardships experienced by the poor. The Mckinsey Global Institute has proposed a new analytical measure, the Empowerment Line, to assess what constitutes a meaningful or minimum standard of living. It is a new metric that takes into account needs of citizens who are above the threshold of the official poverty line but continue to face multiple deprivations. They have also introduced the Access Deprivation Score, to measure number of lives that are marked by a continuous struggle to achieve a modicum of dignity, comfort and security. It is necessary, therefore, that the debate in India about income levels for definition of a poverty line be broadened to consider levels of deprivation which would then include issues such as energy access and other key social determinants such as access to health, drinking water and sanitation.

There are expectedly wide variations between the energy consumption of developed and developing countries and between the rich and poor within countries, with attendant variations in human development (Figure 1.1). Per capita energy consumption in Africa region and India is much less than the world average, which is estimated at 1.8 tonnes of oil equivalent (toe) per person per year.

There is also a direct relationship between fuel use and income. People with low per capita income in developing countries are more dependent on biomass for meeting their energy requirements in the residential, services, industry and transport sectors, and the energy consumption is comparatively low. With increase in incomes, the fuel mix becomes much more diverse and the overall amount of energy consumed is much higher. Figure 1.2 illustrates the relationship.

The International Energy Agency (IEA) has devised an Energy Development Index (EDI) to better understand the role that energy plays in human development. It tracks progress in a country's or region's transition to the use of modern fuels. The EDI is calculated in such a way as to mirror the United Nations Development Programme's (UNDP) HDI and is composed of four indicators: (a) per capita commercial energy consumption, which serves as an indicator of the overall economic development of the country; (b) per capita electricity consumption in

Figure 1.2
The relationship between per capita final energy consumption and income in developing countries

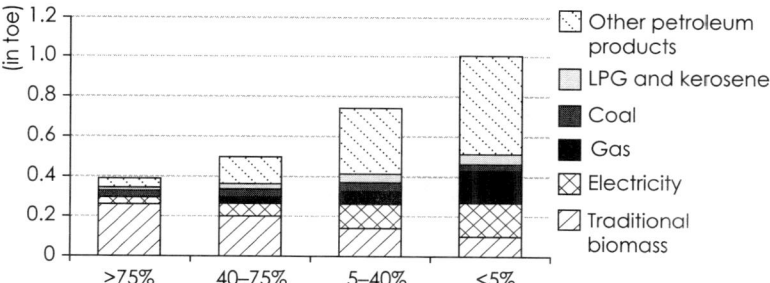

Share of population with an income of less than $2 per day

Source: World Energy Outlook (WEO) 2010.

Figure 1.3
Comparison of the EDI to the HDI

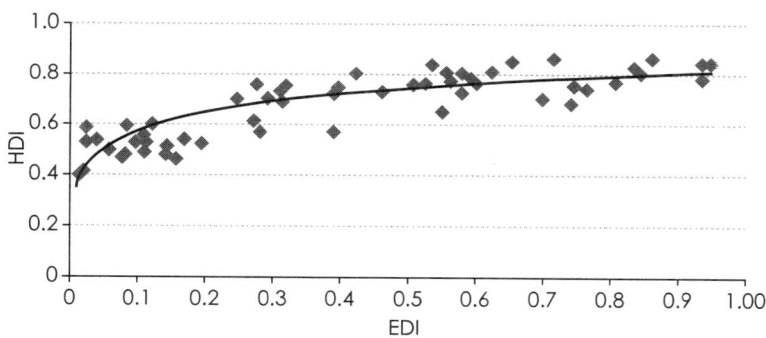

Source: IEA (2010).

the residential sector, which serves as an indicator of the reliability of, and consumer's ability to pay for, electricity services; (c) share of modern fuels in total residential sector energy use which serves as an indicator of the level of access to clean cooking facilities and (d) share of population with access to electricity. Each of these indicators captures a specific aspect of potential energy poverty. It is no surprise that EDI results are strongly co-related with those of the HDI such that lower the energy development level of a country, lower it is on the HDI too (Figure 1.3) (IEA, 2010).

Figure 1.4
Per capita energy consumption and emission in selected regions

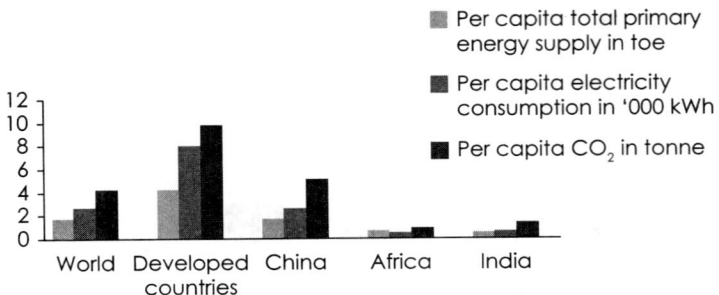

Source: IEA Key Energy Statistics (2011).

The report of the Committee for evolving a Composite Development Index of states has taken into account percentage of households with electricity as primary source of lighting under the category of household amenities.[5] Hopefully, this would result in certain targets for achievements in each indicator if extra funds are given to a particular state.

It would also follow that there would be a direct correlation between energy consumption and emissions levels (Figure 1.4).

This clearly shows that it is the energy poverty of certain countries, which is responsible for their lower emissions, at least in substantial part. (In fact, they have provided some space and time for emissions of other countries to grow.) The environmental or climate change debate must recognise this basic fact. While there could be discussion on how to contain developmental emissions, two approaches must become fundamental and must be recognised. First, every country must be permitted at the minimum to reach the current average global per capita energy consumption multiplied by their current population and all others come down to that level. Second, within a country this calculation should ensure certain minimum levels of energy consumption are factored in separately for people living in rural areas.

It is also almost self-evident that lack of energy access is primarily a rural issue. Although a percentage of the urban population still does not have access to modern energy, particularly in Africa, the problem is acute in the rural context where it is many times greater than the urban areas,

The Importance of Energy Access   9

and where addressing it faces many more challenges. It will also follow that energy access becomes a critical component not only to reduce rural poverty and drudgery but also is one of the fundamental conditions for holistic rural development. Broadening this viewpoint further, many of India's most backward areas, which have the worst HDI indicators and higher indices for multi-dimensional poverty, have large rural populations with very poor levels of energy access. For their overall development, energy access becomes a critical need, perhaps even a precondition.

Notes

1. Affordability is a complex issue where 'cost', 'willingness to pay' and the 'ability to pay' have different connotations. These issues would be discussed later. It also has implications for the economics of supply and related issues of technology, pricing, financing and subsidies. It is not clear what clean means. It could mean that biomass should not be burned for cooking and may be suggesting use of cooking gas, which may not be the right conclusion because of many difficulties. It could also mean electricity, irrespective of the source of generation, rather than lighting from kerosene and power from diesel. But we would like to broaden the definition even further and stress on clean sources of electricity supply and improved cookstoves to be properly recognised here.
2. In 2001, G-8 Task Force on Renewable Energy attempted to quantify what makes subsistence energy that greatly affects people's quality of life. It estimated that to provide basic electricity services for household, community (hospitals and schools) and commercial activity, people need about 50 kWh per person per year. The basket of activities under basic electricity services included the radio a few hours per day, reading at night, consuming a minimal amount of clean water, etc.
3. When Deepak Gupta first visited a village in Bihar in June 2010, to see the functioning and impact of the first Husk Power Systems village electrification, a villager got onto the stage and said that 'for us (read we urbanites) independence came in 1947, for them (read the villagers) it came when the first bulb was lit a few months back'.
4. In a 2005 household survey in rural areas, the authors found 57 per cent of households' energy poor, while 22 per cent were income poor.
5. In May 2013, the Government of India had set up a Committee for evolving a Composite Development Index of states under the Chairmanship of Raghuram Rajan, now the RBI Governor, to study individual states' progress vis-à-vis national progress on defined criteria. The Committee submitted its report in September 2013. The report has developed an index of (under) development, which is an average of 10 sub-components: monthly per capita consumption expenditure, education, health, household amenities, poverty rate, female literacy, per cent of SC–ST population, urbanisation rate, financial inclusion and connectivity. Available at http://www.finmin.nic.in/reports/Report_CompDevState.pdf. Accessed on 22 November 2014.

2
Global Status

Global Scenario

Energy access includes both electricity and cooking energy. However, the energy access policy and literature and discussions reflect relative neglect of the non-electricity household energy access (World Bank, 2011a). Perhaps many policy makers, even in developing countries which are worst affected, are not sufficiently aware of the extent of the other problem. It is essential that both aspects be given importance and actions taken in both areas.

Globally, as many as 1.4 billion people (over 20 per cent of the global population) do not have electricity to light their homes or conduct business (Figure 2.1). Around 2.9 billion people, almost 40 per cent of the global population, rely entirely, or to a large degree, on traditional biomass for cooking and heating (Figure 2.2). These are huge numbers.

It is important to note that while percentages may change over the years, and concerned ministries often rely on them to show progress, the absolute number has not changed appreciably over the last few decades, particularly in rural areas. Even more alarming is the IEA's projection that, if no new policy to alleviate energy poverty is introduced, 1.2 billion people (some 16 per cent of the total world population) would still lack access to electricity in 2030. India's numbers will reduce but will still have about 300 million people without electricity access. Around 2.8 billion people will depend on biomass for meeting cooking energy needs, about 82 per cent in rural areas. The latter number would actually have increased (see Table 2.1).

Global Status 11

Figure 2.1
Share of population (per cent) without access to electricity

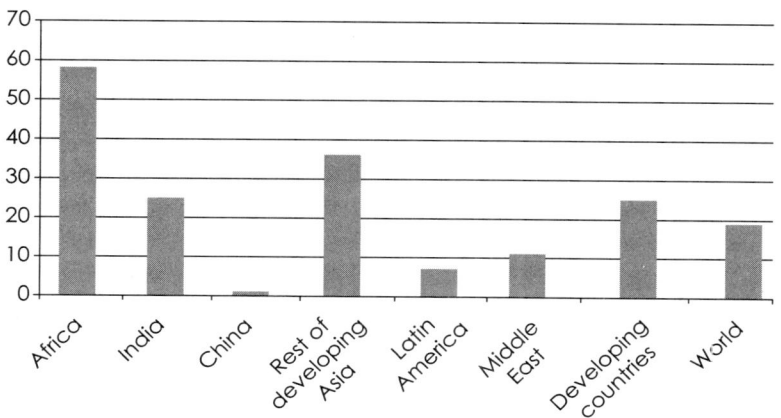

Source: IEA (2011).

Figure 2.2
Share of population (per cent) relying on traditional biomass for cooking

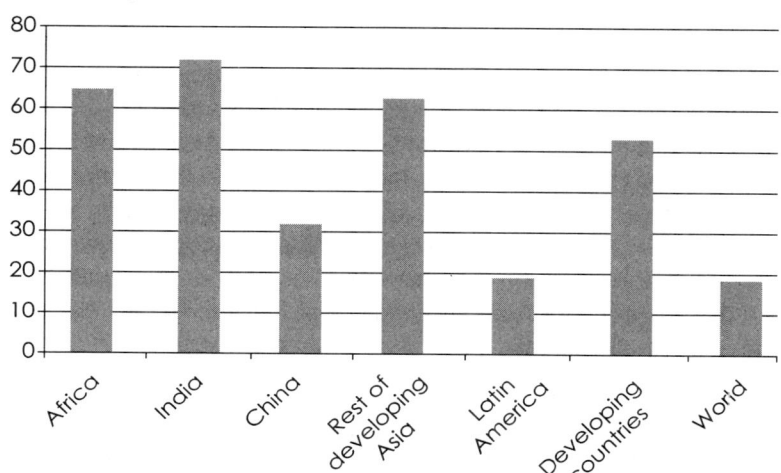

Source: IEA (2011).

Table 2.1
Number of people without access to electricity and relying on traditional use of biomass (in millions)

	Number of People Lacking Access to Electricity			Number of People Relying on Traditional Use of Biomass for Cooking		
	2009	2015	2030	2009	2015	2030
Africa	587	636	654	657	745	922
Sub-Saharan Africa	585	635	652	653	741	918
Developing Asia	799	725	545	1937	1944	1769
China	8	5	0	423	393	280
India	404	389	293	855	863	780
Other Asia	387	331	252	659	688	709
Latin America	31	25	10	85	85	79
Developing countries	1,438	1,404	1,213	2,679	2,774	2,770
World	1,441	1,406	1,213	2,679	2,774	2,770

Source: IEA database and analysis as referred in World Energy Outlook 2011.

Table 2.1 and Figures 2.1 and 2.2 also indicate that the problem is most acute in sub-Saharan Africa and South Asia, with India being the most prominent. Global attention is likely to focus on Africa, and perhaps international funds and efforts would also be directed there. The EU is making efforts there. During his visit to Africa in June 2013, President Obama also declared a US$7 billion 'Power Africa' programme for eight countries over 5 years (Ishofsky, 2013). In some ways, therefore, the Indian challenge appears even more daunting, as it appears that mostly national efforts would be needed. For Indian policy makers, this should be a wake-up call.

International Efforts for Energy Access

This brings us to look at what international efforts have been made to address these challenges. In general, this area has remained neglected and has seen little action globally over the years. This would become

Global Status 13

evident as we briefly look at major international developmental efforts and events.

The Climate Convention and Agenda 21, negotiated in 1992, only agreed subsequently to include the key issue of making energy available to those who do not have it at present in an environmentally sustainable manner. After the Rio Summit on Environment and Development in 1992, the Global Environment Facility (GEF) was established as a global grant-giving fund. At that time, the World Bank was also asked to increase its assistance to renewable energy projects. This led over the next decade to several small World Bank loans for decentralised solar projects, which are discussed in Chapter 5. But nothing substantial happened subsequently. The Millennium Development Goals (MDGs), agreed in 2000, were perhaps the most significant global initiative to give a direction to national and international developmental efforts. They did not include energy as a basic need and ignored its essential role in establishing infrastructure and creating the environment necessary for the eradication of rural poverty. Only now is it being recognised that access to energy services would be critical in achieving all of the MDGs (Modi et al., 2005).

The link between energy services and poverty reduction was also identified by the World Summit on Sustainable Development (WSSD) held in 2002 in the Johannesburg Plan of Implementation, which called for the international community to

> take joint actions and improve efforts to work together at all levels to ... improve access to reliable, affordable, economically viable, socially acceptable and environmentally sound energy services and resources ... through various means, such as enhanced rural electrification and decentralized energy systems, increased use of renewable ... recognizing the specific factors for providing access to the poor.[1]

But once again, no worthwhile global action was forthcoming.

Only in the last few years does energy access seems to have really become a part of the international agenda, and certainly so as regards discussions, etc. are concerned, because concrete actions are still not being seen. Perhaps it coincided with the realisation of the impact traditional energy sources are having on climate change and renewed focus on development of clean energy sources.

In June 2009, the UN Secretary General established an Advisory Group on Energy and Climate Change (AGECC). This was chaired by

Mr Kandeh K. Yumkella, then Director General UNIDO, and now the head of the UN Sustainable Energy for All (SE4ALL) initiative. He has been instrumental in spreading both awareness of and commitment to resolve issues of energy access. This Group produced a Report in June 2010, which made a strong case and appeal for energy access and sustainable energy by stating that it is 'shameful and unacceptable' that a third of humanity has no access to modern energy services and half of humanity has to rely on traditional biomass for meeting their basic needs, including cooking (United Nations, 2010). During the year, the UN developed a new poverty index, which stressed that lack of services such as electricity was a key factor in determining poverty. In September, 2010, as part of the World Energy Outlook 2010, a special report entitled 'Energy poverty: How to make modern energy access universal?' was published jointly by IEA, UNDP and UNIDO.[2] These documents reflect both the enormity of the problem and the recognition of the enormity. They point to this becoming a strategic developmental challenge requiring immediate and focussed attention of both the national governments and the international community.

Given the scale of the daunting challenge ahead, in April 2011, the UN General Assembly declared 2012 as the International Year of Sustainable Energy for All. Further, in 2011, the UN Secretary General launched the Sustainable Energy for All Initiative—SE4ALL, a global energy partnership and campaign aimed at achieving three inter-related goals by 2030: viz., ensuring universal access to modern energy services; doubling the global rate of improvement in energy efficiency; and doubling the share of renewable energy in the global energy mix. The members of the Secretary General's High-Level Group on SE4ALL include very high levels of representation from various fields.[3] SE4ALL convened a global forum in New York in June 2014 where experiences were shared; however, there is still no road map and firmed up finances. Mr Kandeh Yumkella, Chief Executive Officer, SE4ALL summing up the discussions said

> SE4ALL is about bringing partners together, looking holistically at energy issues and giving everyone a voice ... the sustainable energy movement has just started, and there is a need to decide the long-term institutional framework for sustainable energy partnerships: what will be its legal form, and how will it be financed and supported. It's a long journey, and we have just taken the first step.[4]

Global Status   15

The first steps need to be taken quickly and firmly.

United Nations Conference on Sustainable Development (UNCSD), also referred to as 'Rio+20' was held at Rio de Janeiro, Brazil, in June 2012. There was hope of major breakthroughs for the energy access agenda. However, while the determination to make sustainable energy a reality was re-affirmed, no firm commitments were made.[5] One could be disappointed at the lack of a firm push or direction in so far as energy access was specifically concerned. The UN Secretary General's SE4ALL initiative was also not formally endorsed.

In the same year, the UN General Assembly decided to keep the emphasis on the importance of sustainable energy by declaring 2014–24 the Decade of Sustainable Energy for All.

This Declaration, however, did not define any tangible targets, and only called upon governments, as has been happening in previous conferences, to take further action to mobilise the provision of financial resources, technology transfer on mutually agreed terms, capacity-building and the diffusion of new and existing environmentally sound technologies to developing countries and countries with economies in transition, as set out in the 'Johannesburg Plan of Implementation'.

The annual Vienna Energy Forum meets have become a platform to bring together high-level dignitaries, policy makers and strategic energy partners to engage in dialogue on the centrality of sustainable energy and energy access in the post-2015 development agenda. The Vienna Energy Forum 2013 held during 28–30 May recalled the energy goals for 2030 stating that they will be monitored closely.[6] It also stated that global transformation towards sustainable energy is necessary and the window of opportunity for action is now and that the business-as-usual approach will not suffice. Energy must be fully integrated into the global post-2015 development agenda.

There nevertheless seems to be a developing consensus and recognition that energy access requires an altogether different set of strategies and actions. Basket of generic suggestions that are being deliberated include (a) developing country governments should commit to expanding access to modern energy services by making it a national development priority—first the political commitment and then it should translate into action plans backed by adequate financial and technical resources; (b) strategies for rural electrification should be based on decentralised power generation;

(c) attention to differentiated energy needs with wide range of energy technology options; (d) promoting efficient use of biomass through the use of improved cooking stoves; (e) active involvement of developed countries for ensuring availability of energy access technologies, partnering in innovation and also for enhancing the technological capability; (f) considering energy access technologies as global public good and work for an international commitment in terms of finances and technology package; (g) develop an enabling environment and framework for generating resources for energy access projects.

But the HOW is really not clear nor has there been a road map prepared. Numerous international conferences have being held on this subject, and these are held regularly.[7] Considering the amount of discussions, and their frequency, and the importance of energy access being repeatedly emphasised, and exhortations made, it is unfortunate that no specific global programme has been launched, or even been prepared. Over the years, there have been sporadic efforts and pilot projects in Africa and India and elsewhere in different parts of the world, largely funded by the World Bank or other multi-lateral and bilateral aid agencies. These have been largely uncoordinated, and have not contributed either to scale or replicability. There have been many studies. But the challenge continues to be largely unaddressed in terms of preparation of a road map and determined and substantial practical steps being taken.[8]

Why should this be so? There are crucial differences of approaches and priorities, which impact action. While developing countries are focussed on securing both energy for development and energy to meet basic human needs, the developed countries are primarily concerned with increasing domestic energy security and decarbonising their energy mix, and their concern is also with reduction of global carbon emissions. IEA estimates suggest that achieving universal energy access by 2030 would increase global electricity generation by 2.5 per cent, which in turn will result in increase in fossil fuels demand by 0.8 per cent and CO_2 emissions by 0.7 per cent only (IEA, 2011). It is rightly recognised that a large part of the solution will come through decentralised renewable energy systems (Table 2.2). In fact,

> decreasing costs and improving reliability have led off grid renewable energy technologies to become the most cost effective option for electrification in most rural areas. In recognition of the critical role renewable

Table 2.2
Generation requirement for universal electricity access, 2030 (TWh)

	On-grid	Mini-grid	Isolated Off-grid	Total
Africa	196	187	80	463
Sub-Saharan Africa	195	187	80	462
Developing Asia	173	206	88	468
China	1	1	0	2
India	85	112	48	245
Other Asia	87	94	40	221
Latin America	6	3	1	10
Developing countries	379	399	171	949
World	380	400	172	952

Source: IEA (2010).

energy will play in achieving the goal of universal energy access it must be integrated into national rural electrification strategies.[9]

These figures showing marginally increased 'poverty emissions', while leading to improvement of quality of life and inclusive growth for the rural poor can also be considered non-consequential if the entire gamut of increase in fossil fuel consumption and consequent emissions are considered.

Both energy efficiency and electricity generation by renewable sources lead directly to significantly lesser emissions. That is perhaps why, in most of the Reports and Conferences, energy access gets clubbed with grid electricity mix and energy efficiency, which dilutes the emphasis on energy access alone. Further, they tend to become the focus of 'sustainable energy'. Discussions also go back to the need for traditional solutions through strengthening of the grid and increasing supply of electricity.[10]

Actually, renewable energy solutions for energy access need to be recognised as an essential part of the global emission reduction story, because a minor increase in emissions is actually replacing much larger alternative emissions if done the conventional way and consequently considered as a significant global public good.

The above description while indicating the pressing need for global action to achieve energy access also tries to rationalise its absence in any meaningful way. In this context, therefore, and taking all these things together, it must be recognised, in more ways than one, that ensuring energy access will primarily remain a developing countries' challenge, and one perhaps that may have to be largely addressed by the developing countries themselves. It is no surprise, therefore, that even though there is recognition of the great need of international finance, not much has actually been forthcoming, and one fears that this lack of availability will continue, especially in the volumes required. Consequently there could be continued international neglect of large-scale action on energy access, especially through decentralised renewable energy–based systems.

India has been making efforts to address these issues, though the progress is limited and much more needs to be done. Let us now see what the electricity status of the country is, which suggests that renewable energy solutions are not only desirable but are going to be necessary.

Notes

1. Johannesburg Plan of Implementation. Available at http://www.un.org/esa/sustdev/documents/WSSD_POI_PD/English/WSSD_PlanImpl.pdf. Accessed on 22 November 2014.
2. This was published in the form of special early excerpt of the World Energy Outlook 2010 for the UN General Assembly on the Millennium Development Goals. Available at http://www.unido.org/fileadmin/user_media/Services/Energy_and_Climate_Change/Renewable_Energy/Publications/weo2010_poverty.pdf. Accessed on 22 November 2014.
3. These include Chairman, Bank of America; Minister of New and Renewable Energy, India; Director-General, OPEC Fund for International Development; Director General, International Renewable Energy Agency; President, World Business Council for Sustainable Development; Secretary of Energy, the United States; UNDP Administrator; Director-General, Russian Energy Agency; Special Envoy for Climate Change, World Bank; Executive Director, United Nations Environment Programme; and Chairman, China Development Bank. High Level Group on Sustainable Energy for All. Available at http://www.un.org/wcm/content/site/sustainableenergyforall/home/members. Accessed on 22 November 2014.
4. Sustainable Energy for All Forum: 4–6 June 2014. Available at http://www.se4all.org/wp-content/uploads/2014/06/Summary_of_the_SE4ALL_Forum.pdf. Accessed on 22 November 2014.

Global Status   19

5. The Future We Want, UN General Assembly, 24 July 2012. Available at http://daccess-dds-ny.un.org/doc/UNDOC/LTD/N12/436/88/PDF/N1243688.pdf?OpenElement. Accessed on 20 September 2014.
6. Summary Report Vienna Energy Forum 2013. Available at http://www.unido.org/fileadmin/user_media/Services/Energy_and_Climate_Change/Renewable_Energy/VEF_2013/Summary_Report_of_Vienna_Energy_Forum_July_4_2013_final.pdf. Accessed on 22 November 2014.
7. India hosted an International Seminar on Energy Access in October 2012. The seminar was attended by Ministers and Government Representatives from 42 countries and it resolved to continue to make concerted efforts for achieving universal access. Abu Dhabi hosts the Future Energy Summit every January. IRENA organised a conference in Abu Dhabi in 2011 and also in Ghana in 2012. UNIDO hosts the Annual Energy Summits in Vienna. There have been several other discussions organised. In each of these, the importance of the subject is stressed.
8. This has been Deepak Gupta's argument in the many Conferences that he attended as Secretary, MNRE, or as an expert, and spoken in.
9. Director General, IRENA, in his foreword to the proceedings of the IOREC 2012. Available at http://www.irena.org/DocumentDownloads/Publications/IOREC_Key%20Findings%20and%20Recommendations.pdf. Accessed on 22 November 2014.
10. Deepak Gupta has been witness to the emphasis on these issues in such discussions.

3
India Electricity Status

Background

India's commercial energy use has increased 21 times and the power generation capacity has gone up by over 100 times during the past 60 years. In 2011–12, the total commercial energy supply (TCES) was 537 mtoe,[1] as compared to about 26 mtoe in 1953–54.[2] The installed capacity at the beginning of the First Five Year Plan was 2.3 giga watt (GW), which increased to 245 GW by March 2014.[3] While this would appear to be a tremendous achievement, we must see this with respect to five important qualifiers in respect to the overall performance of the power sector. First, a substantial part of this generating capacity, particularly thermal, has actually come about only in the last few years, because of a special and much-needed policy thrust. Over 40 per cent of the total thermal capacity of 168 GW as of 31 March 2014 was added between 2006 and 2014. In 2000, our total power-generating capacity was only 112 GW. In comparison, China grew from a capacity of 325 GW in 2000 to 1,230 GW by March 2013. However, the last couple of years have seen a significant break in momentum with many long-term implications. Second, while supporting around 16 per cent of the world population, India's share in world energy use and electricity consumption is only 4.2 per cent and 3.5 per cent, respectively.[4] In the international context, however, when fears of India's future emissions are projected, this figure is shown with the argument that India is already the world's fifth largest energy consumer. It is, therefore,

Figure 3.1
India has low per capita energy, electricity consumption and CO_2 emission

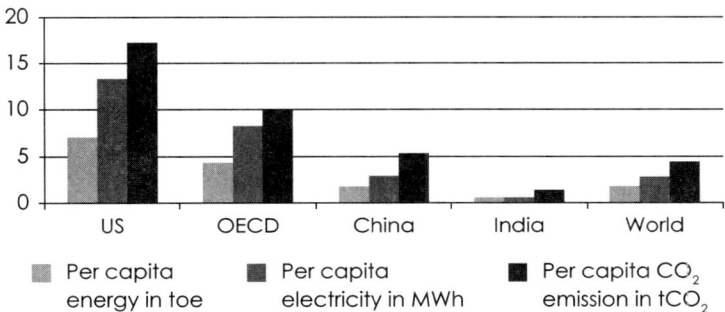

Source: IEA Key World Energy Statistics, 2012.

important to note that the current per capita energy use at around 0.58 toe is far below that of industrialised countries, and, more importantly, only a third of the world average (Figure 3.1).[5] Incidentally, even with the huge growth in energy consumption projected over the next two decades, per capita consumption in 2032 would only be 74 per cent of the world average in 2003.[6] This reflects a huge latent or potential demand. Third, there has been a galloping growth in demand caused by high rates of growth, rapid urbanisation and an increasing appetite for consumer goods from a growing middle class. But, in spite of significant growth in electricity generation over the years, significant shortage of power continues to exist primarily on account of growth in demand outstripping growth in supply. Energy deficit levels have remained consistently high in recent years with supply trailing requirement by an estimated 8–10 per cent,[7] rightly described as 'astronomical'.[8] While this figure is generally accepted for purposes of analysis and policy, there is also a need to question the level of deficit itself, because the latent demand appears never to have been taken into account. The dramatic grid failure of 2012 and the power shortage crisis in Delhi and neighbouring areas during the summer in 2014 were symptoms of the larger problem of India's energy sector not being able to keep pace with the demand. Fourth, the distribution sector continues to show incomplete infrastructure, continuing inefficiencies and substantial transmission losses. Moreover, there are adverse consequences for Discoms if they are unable to charge market rates of power or get tariff above the cost of generation. Finally, as already stated, 300 million people estimated to be without electricity access.

Current Electricity Mix

The electricity mix of India's grid connected power generation installed capacity (245 GW) as of March 2014 and the share of renewables in total electricity generated is given in Figures 3.2 and 3.3, respectively.[9] The most important feature is the dominance of coal-based power production.

Figure 3.2
Percentage share of renewable power in the electricity installed capacity

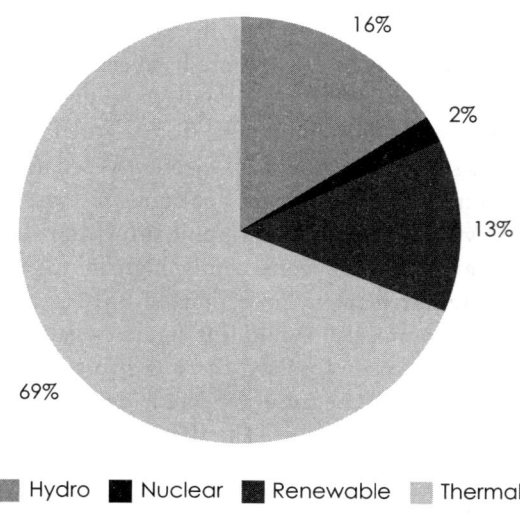

■ Hydro ■ Nuclear ■ Renewable ▨ Thermal

Source: Central Electricity Authority.[10]

Figure 3.3
Percentage share of renewable energy in electricity mix

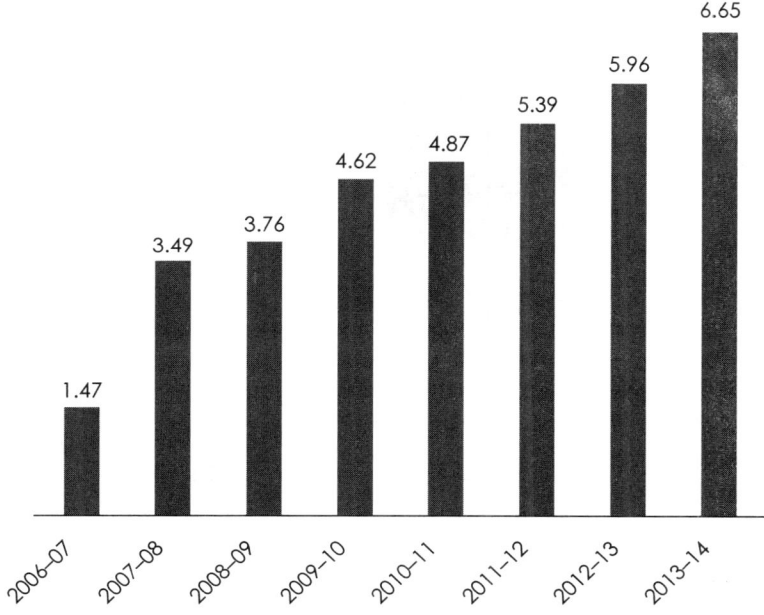

Source: Central Electricity Authority.[11]

The break-up of renewables installed capacity by source is given in Figure 3.4. The total source-wise electricity generated for the period April 2013–March 2014 is given in Figure 3.5.

Future Projections

But what are the portents for the future? The Integrated Energy Policy (IEP) Report, published by the Planning Commission in 2006, has been the main study, which has forecast electricity demand and supply in India. It forecast that power-generating capacities of 778 and 960 GW would be needed with 8 per cent and 9 per cent growth by 2032, respectively. These demand estimates are likely to be lower

Figure 3.4
Renewable power installed capacity

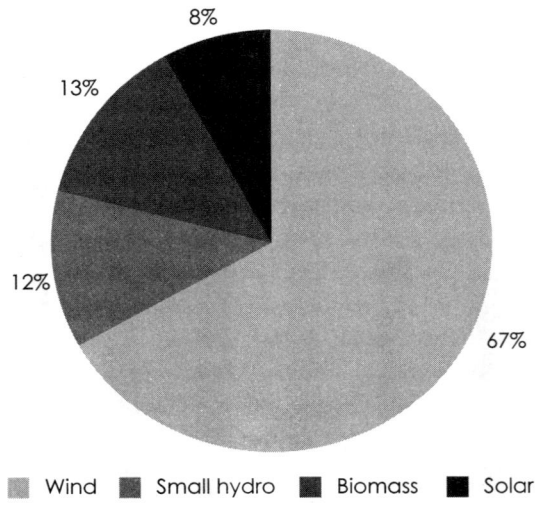

Source: Ministry of New and Renewable Energy.

Figure 3.5
Technology-wise electricity mix 2013–14

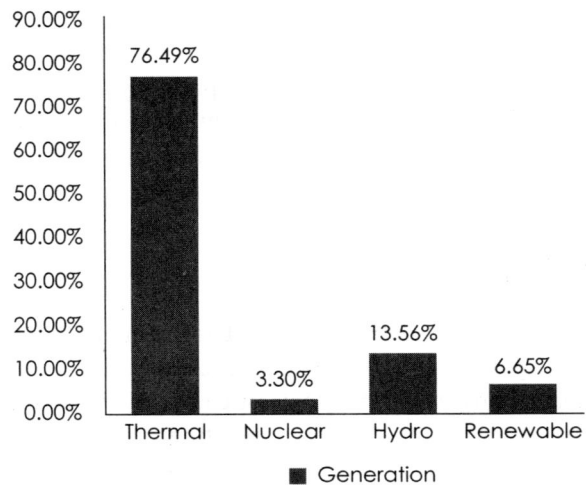

Source: Ministry of Power and Ministry of New and Renewable Energy.

because of commitments made to reduce energy intensity of GDP and likely lower growth rates. Estimates of the possible sources of supply were also made. Projections on the likely electricity generation by different fuel supplies by the end of 2030 have also been made in the 12th Five Year Plan document, though these have been given in percentage terms not in absolute numbers. There have been some significant changes in the projections made in the latest Report of the Expert Group for low carbon strategies.[12] Total capacity has been reduced as well as that of conventional sources while that of renewable energy sources has been considerably increased. The projections of demand and supply of power are very important because of their serious implications for electricity access in India, now and in the foreseeable future. Therefore, it is necessary to clearly understand the likely future developments.

Conventional Sources

A decade ago, there was great confidence in the power future of India. The Electricity Act had provided the framework. Large capacities for thermal and gas were encouraged, and planned for large hydro. Coal was being allocated to public and private sector power-generating companies in captive mode. The civil nuclear deal with the US seemed to have removed critical obstacles for the growth of large-scale nuclear power. Unfortunately, however, all these sources started facing increasingly serious general problems of forest and environmental clearances; production difficulties; infrastructural and logistical bottlenecks; constraints of financial investments and bank credit; popular discontent with nuclear, increasing costs of imports, etc. as well as some specific issues related to each sector. Although these are being addressed, it appears that there will be substantial shortfalls in setting up the projected capacities, both in the short and the long term. This has been increasingly recognised in discussions and articles/reports. Main issues are briefly summarised.

Coal

India is the world's third largest consumer of energy from coal accounting for around 58 per cent of the country's total commercial energy consumption. But serious problems are emerging regarding production, transportation, import, costs, emissions and pollution because of which there is considerable uncertainty of what will happen in the future.

- India nationalised its coal mining industry in 1976. Since 1981, coal consumption had increased at an annual rate of 5.6 per cent, which was being supplied by Coal India. Realising that the public sector coal companies would not be in a position to meet the increasing demand for coal, the government initiated the coal block allocation process in 2004 to public and private firms wanting to set up power plants.[13] However, many companies did not take serious action with the result that only a few of these blocks have become operational, and some are being cancelled. This is going to substantially delay production from new mines and the costs may also go up if and when they do produce.
- A substantial portion of our reserves are underground, which requires India to go beyond open cast mining and exploit this potential, particularly when forest clearances are difficult to come by. Perhaps professional mining companies may have to be involved. A policy framework for this has yet to be developed. This coal will also be costlier.
- There is a general perception that India's coal reserves are plentiful and could last over 100 years. However, some experts have cautioned against the 'myth of abundance'. The IEP warned that 'large estimates of total coal resources give a false sense of security'. The 11th Five Year Plan indicated extractable coal reserves will run out in 45 years if coal production continues to grow at 5 per cent (Ahn and Dagmar, 2012).
- The coal gap is currently about 67 million tonnes (MT). Since coal demand is increasing by 10 per cent and production only by 5–6 per cent the gap is likely to increase by 2017 to 178 million tonnes (MT), which has been identified in the 12th Plan as import requirement in that year.[14] The Low Carbon Expert Group has projected a requirement of 1,568 MT of coal by 2032 in business as usual scenario reducing to 1,278 MT if most of the new thermal power capacity comes up as super critical plants. This would mean ever more imports of coal. Cost of import, already a problem, will become a critical factor, while adding considerably to the current account deficit burden and raising concerns of energy security.

India Electricity Status 27

- There are increasing problems of logistics, including carrying capacity of the Railways. Discussions with Railway and Coal officials indicate that the absence of, and delay in, establishing special coal corridors could become a serious connectivity constraint in the future. For the Railways, there are serious fund constraints and none of the projects have even got environmental clearances as yet. Transportation of imported coal will also be a problem. The four major ports that handle coal, having a capacity of about 63 MT, are on the Eastern coast. Most of the power plants will be in the hinterland and some in Western India. Even the handling capacity in the ports could be a problem.
- As the years go by, and as production increases, the environmental problems, already critical, are likely to increase. Most of our coal lies in tribal and forest areas. Getting 'green' clearances to mine has been, and will continue to be, a problem. Pressures to contain or reduce carbon emissions in India will also become constraining factors. It is estimated that India's carbon emissions are likely to grow by 43 per cent between 2011 and 2020.
- The actual generation from thermal plants has shown capacity utilisation of only 70 per cent in the last couple of years because of the problems of supply of coal, whether domestic or imported. This trend is likely to continue. All this has also caused problems for the lenders with implications for future financing of power projects.[15] In fact, the burdened banking sector with a portfolio of non-performing loans to power plants has added a new dimension to the many problems of the power sector.

With the present thermal capacity being 132 GW, it is difficult to see this doubled by 2032, or meet the coal requirements for this capacity, should it be set up, even though approvals for a much larger capacity may have been given.[16] All this, and increasing cost of coal leading to higher generation costs, will have serious consequences for rural supplies of coal-based power.

Currently, earnest efforts are being made to address all these problems but it will be a difficult path.

Natural Gas

The share of natural gas in India's commercial energy mix was around 5.6 per cent in 1991, which had risen to around 8 per cent by 2010. With a

12.2 per cent annual growth rate, it is one of the fastest growing sources of energy. Around 45 per cent of natural gas is used for power generation. In 2010, 46.5 billion cubic metres (bcm) of the 47.2 bcm of gas consumed in India was sourced domestically. India imported gas for the first time in 2004 in the form of liquefied natural gas (LNG). Over the years, the share of import is expected to increase because projections of domestic gas supply have gone quite awry. Imports were 28 per cent in 2011–12. They are likely to be 70 per cent by 2017. Since 2005, domestic natural gas output has increased by barely 1 per cent annually. In 2009, the Petroleum Ministry projected availability of 186.6 mcm per day with this level being maintained roughly for the next few years. The current projection is 105 mcm per day for this year going up to 139 mcm in 2015–16. Consequently, about 15,000 MW of the 16,000 MW capacity set up has been stranded and supplies to the 6,000 MW capacity under construction are also doubtful for some time. All these plants were set up following a comfort letter from the Ministry of Petroleum assuring supply of gas for plants set up during the 12th Five Year Plan. Though there could be increase in domestic exploration and supplies, as also availability through imports, there would be physical constraints and increasing costs, which would be substantial even with the lower price increase decided by the new government. Issues of competitive uses and prioritisation of other sectors and the struggle with policy complications regarding the price, now and in the future, will continue. Thus, there will be no assurance on availability and there will be no price predictability. There simply will not be any financing in such a situation. All of these suggest that gas power-generating capacity may not be more than 30 GW as against the 70 GW projected in the IEP Report. Even this may be doubtful. Actually, a view needs to be taken immediately whether at all there should be any further addition to gas-based power plants except as spinning reserve or for evening peaks, and in that case, how would they be set up and what would be the tariff. This is in the context of likely large-scale grid penetration of renewable power a couple of decades hence. This issue needs to be addressed.

Large Hydro

India is endowed with economically exploitable and viable hydro potential assessed to be about 84 GW at 60 per cent load factor (148.7 GW installed capacity). In addition, 15 GW in terms of installed capacity from small hydro, mini and micro hydro schemes has been assessed. Also, 56 sites for pumped storage schemes with an aggregate installed capacity of 94 GW have been identified (Government of India, 2008a). In 2010, India was the seventh largest consumer of hydroelectricity in the world, with this source accounting for 3.3 per cent of the world's hydroelectricity generation (IEA, 2012). However, large hydro has been constrained by environmental factors. While the Chinese are going ahead at breakneck speed to use the Himalayan waters, we have a variety of problems, which are likely to prevent them from being harnessed as planned. Last year's rainfall deluge in Uttarakhand, while damaging two ongoing large hydro plants, also is a pointer to future constraints. Two projects were abandoned some time back and in August 2013, the Supreme Court put a restriction on future projects. The projects in Arunachal Pradesh are taking a lot of time.[17] Hydro sector will be very important for India, not only because it is sustainable, safe, clean and reliable, but also because it is a good source of peaking power and relatively cheaper, and we need both the water and the power. Therefore, focussed action is needed. We mostly agree with the 10-point plan suggested for this purpose (Chatterjee, 2013). Pump storage projects would also be needed for balancing loads. However, with our best efforts, we may not be able to reach more than 60 GW as against the 150 GW projected by IEP.

Nuclear

India's current nuclear capacity is 5.78 GW. Anti-nuclear protests have delayed commissioning of the reactor units of Kudankulam Nuclear power Project 1&2 (2×1000 MWe) set up in technical collaboration with Russian Federation.[18] Capacity under construction is soon expected to be about 4,800 MWe.[19] At some places, pre-project activities are starting. Approximately 10,500 MW with domestic resources and 8,000 MW with foreign investment are on the anvil.[20]

There are several issues of concern.[21] The most important and worrying is that capacity forecast has consistently been much higher than its actual creation. In 1964, Dr Homi Bhabha had stated: 'there is little doubt that before the end of the century, atomic energy would be producing a substantial part of the power in India, and therefore practically all the addition to our power generation will come from it at that time'. The then Atomic Energy Commission (AEC) Chairman, Dr Sethna, quantified this in 1972, forecasting that India would have 43 GW of nuclear-generating capacity by 2000. However, we reached only 2,720 MW. In 2004, DAE analysts projected that the total capacity would become a huge 275 GW by 2052, and in 2008, the then Chairman AEC, Dr Kakodkar, stated that India would face a shortfall of 412 GW in 2050, which would be practically wiped out by nuclear energy. These were very optimistic forecasts. There has been a much-needed moderation in the latest capacity forecast of 10,080 MW by 2017 and 17,080 MW by 2022. Therefore, to expect >60,000 MW by 2032, as projected by IEP, seems quite unrealistic. Besides there is increasing opposition to nuclear power stations and all proposed sites are facing problems from local populations. After Fukushima, this has become an added dimension of difficulty. The World Nuclear Industry Report 2013 says that nuclear power reached its zenith 20 years ago when its share of global electricity generation was 17 per cent, now it is only 11 per cent.[22] Immediate prospects, therefore, are not good. There are also issues related to the vexed question of liabilities, and serious differences have arisen with likely foreign investors although attempts to resolve these are being made. Maybe in the long term there will be technological advancements, which would regenerate this source of electricity. India needs power, and nuclear power development will be important in the long term. So we must try and maximise it.[23]

Renewables

Inevitably, therefore, greater hope is being placed on renewable energy, especially as in the last few years there has been tremendous growth in renewables capacity, up from 3 per cent of total capacity in 2006 to 13 per cent in 2013, growing over 20 per cent on the average annually over the last decade. However, in terms of electricity supply, the percentage has been somewhat limited, rising from 1.47 per cent in 2006 to around 6.65 per cent in 2013–14. The National Action Plan for Climate Change (NAPCC), announced in 2008, targeted this to become 15 per cent by 2020 itself. But this is too ambitious and simply cannot be achieved. The 12th Plan document talked more realistically of installed capacity share of renewable energy being 33 per cent by 2030, and generation share 16 per cent. The Report of the Low Carbon Expert

Group raises this to 18 per cent with both wind and solar capacities being greater than 100 GW each. The generation percentage is indeed possible to be achieved, but not because of the huge capacity of renewables that will be set up, but much more importantly, because both conventional and renewable capacities may not reach the numbers proposed in the latest documents.

Both wind and solar are likely to grow rapidly but these capacities are very ambitious. In fact, there is talk of even higher capacities.[24] This has led to many pronouncements of developing this kind of power with ultra large solar plants.

Advancements in wind technology have allowed higher masts leading to better harvesting of the lower wind speeds in India.[25] Greater efficiency of solar cells and drastic reduction in costs makes achievement of grid parity possible in a few years from now. However, land and environmental issues (Gadgil et al., 2012), evacuation infrastructure, grid capacity, scheduling and balancing requirements, and even the abilities of the utilities to procure renewable power, could be limitations for both wind and solar. We are presuming annual averages of about 2 GW+ for wind and 2 GW for solar. These appear to be more realistic estimates, though solar could increase substantially. For the future, biomass power needs to be given importance. Since there are many problems for supply of biomass, including both availability and cost, it is essential that we promote decentralised biomass generation, preferably through dedicated plantations of fast growing bamboo and tree species on wasteland and degraded forest land. This could lead to possible capacity of 20 GW in course of time. This would also be useful for base power supply in the future and for balancing requirements and, as we shall see, could be for decentralised generation at the distribution level. We could, therefore, get about 150–170 GW by 2032—wind 60–70 GW; solar 60–80 GW; small hydro 10 GW and biomass 10 GW. Development of this renewable energy capacity in the future and beyond, however, would require, amongst other things, development of an integrated transmission infrastructure; proper location mapping, reservation of land and development of solar parks for a future capacity of 100 GW; development of policy for energy plantations and possible development of spinning reserve. Simple projection of capacities will just not be good enough. And we must be cautious not to over-project, as has been the case in conventional sources, particularly nuclear. In fact, the time has come for a real integrated electricity policy because we would require this in terms of transmission infrastructure, transfer of power from generation points to load, grid management (including flexibility and base loads and therefore integration with gas plants), tariff during different times of day to manage peak loads, etc. The current experience of Germany of integrating renewable energy should provide useful lessons. Actions have to be initiated now.

Implications

Predicting the future in times of uncertainty and changing dynamics is always a hazardous task. It is indeed a wonder how different projections over the years have turned out to be so unrealistic. The problem is that such overly optimistic projections provide a comfort level regarding likely future availability of power, which tends to foreclose other necessary alternatives that need to be pursued. The above analysis suggests that power capacity additions by 2030–32 as projected in the IEP report or in the 12th Five Year Plan document are going to be a big challenge, and in all likelihood, India will end up having not only reduced installed capacities but even more reduced electricity generation. After a detailed review, one analysis has reached a sobering conclusion: 'The bottom line about India's future in the next few years ... is set to get worse, not better. The supply—demand gap for each type of (conventional) energy is set to grow' (Tranum, 2013).

This is so at a time of galloping demand. According to IEA, energy demand in India up to 2035 will show the highest growth rate in the world. Energy consumption in India rose by half between 2001 and 2010 and is expected to swell over 90 per cent from 2011 to 2020 maintaining a path of 'steep ascent', but its generating capacity will grow by less than 70 per cent, because of which India's energy supply will be constantly playing catch up to the demands of its burgeoning economy (Adams, 2014). Continued shortfalls are inevitable especially after factoring in power losses and poor capacity utilisation leading to outages which may 'continue to stall the engines of industry and commerce'. It is obvious that electricity supply would be central to India's economic development and growth and today presents quite evidently the most formidable policy and structural challenge for the country. It also follows that the increase in demand is largely in a rapidly urbanising India, which, along with the manufacturing sector, will continue to absorb all the additional supplies.[26] On the one hand, a continuing situation of supply shortage necessitates immediate adoption of a range of energy efficiency, demand management and conservation strategies, particularly in urban areas.[27] On the other, it raises the critical question— how will we supply power to the rural areas better or more than what we are providing today? Clearly, renewables are set to play a major role in the developing future for grid power, but that power will also be absorbed largely elsewhere, and there will be limitations on what is doable. So, in a situation of supply side constraints, an increasing percentage of renewable grid power generation will not by itself solve the problem of electricity access in rural areas. If a large portion of it is solar, it does not help evening supply, when it is most needed. Moreover, wind power will generally be concentrated in the South and West of India while the rural electricity deficit is more in the East. Besides wind power is heavily concentrated in the monsoon months. Therefore, even as we concentrate on development of renewable grid power we simply cannot afford to ignore its off-grid role, particularly for energy access.

Capacity creation and generation are not the only problems as far as conventional sources are concerned. There are inter-related issues of cost of power, carrying cost, transmission losses, tariffs and the financial condition of the utilities, which we need to consider, because they all have important implications for supply of grid electricity to rural areas.

Cost of Power

Cost of Generation

The cost of electricity from coal-based thermal power plant at Mauda (Maharashtra) being set up by NTPC has been computed at ₹4.85 per kWh for 2014. In this case the coal price has been taken as ₹3.76 per kg, including the coal transportation charges.[28] Whether for existing plants or new, the cost of coal-based power is going to be a problem in the future, particularly for imported coal. There is already evidence of likely, and significant, increase in cost of power from gas. Even though the global gas availability has increased dramatically and consequently imported prices may come down after a few years, the fact is that power from gas will be expensive, probably over ₹10 a unit and increasing. The delays and costs of exploiting hydro power in Arunachal Pradesh and other states may similarly increase hydro unit electricity costs too. Even the future cost of nuclear power is going to be a problem, although there is no real information about the cost of this power.[29] What then of the future cost from the different conventional power sources, individually or together as a combination? In fact, one can sense an irony that the expensive renewable may soon not only have grid parity but may become significantly cheaper than the others as time goes on. This has implications for the nature of our tariff structure and the impact on utilities.

Cost of Supply

There is also a transmission cost that has to be factored in. The costs of setting up transmission and distribution capacities are high and are accompanied by significantly higher losses. 'No one really knows how much is consumed by the agricultural sector and until we know, we won't know how much is being lost', said Dr Promode Deo, then Central Electricity Regulatory Commission (CERC) Chairman, speaking in April 2012 (Tranum, 2013, p. 70). Grid extension costs are primarily distance dependent and break even 'distance' is related to demand. On average, the cost of grid extension increases the cost of electricity supply by approximately ₹1 per kWh/km (Cust et al., 2007). Besides there are transmission losses, which in India are substantial, although average losses came down from 34 per cent in 2001–02 to 24 per cent in 2010–11.[30] The cost of supply to rural areas, therefore, is very high and is multi-dimensional in nature. These problems are not going to away easily or soon. The Shunglu Committee (2011) which had studied this subject in depth has called for immediate radical action stating that soft options are no longer available.[31]

India Electricity Status 35

> **Box 3.1 Cost of electricity supply in rural areas**
>
> The actual cost of supply to rural households is significantly higher than 'average cost of supply' as stated in SREC Orders, which compute this as 'average power purchase cost' + 'average transmission cost/losses' of STU + 'average distribution cost/loss for Discom. Supplies from DDG facility based on RETs should be deemed as displacement cost of 'peak power' since they are mainly during peak load period of Indian grids, i.e. 6–10 p.m. Consequently cost of power should be taken as ₹6/kwh (being the median value of peak power being purchased by STU). Transmission costs (O&M charges + transmission losses) would have to be added. KERC Order for FY 2012–13 approved KPTCL O&M costs of ₹0.56/kWh and 3.96 per cent transmission losses. Based on cost of Power Purchase of ₹6/kwh, the total cost of supply by STU would work out to be ₹0.9/kWh. Consequently, the 'displaced' cost of supply by STU to Discom, between 6 and 10 p.m., would be ₹6.9/kWh.
>
> Discom distribution costs (O&M + losses) would have to be computed in context of supplies to rural households through 415 volts line with low load factor. As per KERC order for CESC for FY 2012–13, the average O&M cost approved was ₹1.05/kWh. Assuming 50 per cent load factor for 415 volts feeder in rural networks, the extrapolated cost for rural household with consumption < 30 kWh/month would be ₹1.5/kWh. Likewise, assuming 10 km 333/11 kV transformer + 5 km 11 kV line + 63 kVA, 11 kV/415 volt transformer + 1 km 415 volt line, the distribution losses for such feeders would amount to 25 per cent or ₹1.73 per kWh (based on ₹6.90/kWh displaced cost of STU supply of power during peak load period). Consequently, the total cost of supply by Discom to rural households would be ₹10.13/kWh. The above figures are meant to be representative rather than a tariff computation.
>
> The key point is that Electricity Supplies from new DDG capacity in rural networks need to be compared with 'displaced cost of supply' through rural electric networks with low load factor and power being sourced during 'peak load period' ... rather than SERC tariffs for domestic consumers (which are invariably subsidised). If this is done, DDG based on RETs will be the preferred solution.
>
> *Source:* BIOENERGY Vision Statement for India. Paper prepared by CREWA, an NGO set up to promote Energy Access, for World Renewable Energy Technology Congress, Delhi, 2013.

Tariff Issues

The problem has been compounded by the supply of power at relatively cheap rates to the rural areas. This approach shows no sign of changing. The agricultural sector consumed 23 per cent of the total power sold in

2011 but got only 7 per cent of the revenues realised from sales (World Bank, 2014), and in some states it is even less than 1 per cent. In Maharashtra the average cost of power is ₹5.56 a unit but farmers pay only 50 paise. Agricultural subsidies accounted for 64 per cent of total subsidies of ₹71,000 crore of states, having risen by 37 per cent between 2007–08 and 2011–12. Moreover, receivables of ₹48,589 crore were uncovered meaning Discoms did not receive the committed subsidies.[32] In this scenario, the more power is given to rural areas the more the utilities lose—the virtue of increased sales turns into the vice of greater losses! (Barnes and Sen, 2002). Therefore, they have no incentive to supply, quite apart from their capacity to do so. There was an implicit understanding that rural supply losses would be subsidised by other tariffs and budgetary support. However, with generation costs going up, urban tariff of various categories themselves becoming controversial and rising and the mounting problems of both budget and fiscal deficits of state governments, both these options are becoming more and more difficult, leading utilities dependent upon operational subsidies becoming critically vulnerable.[33]

Traditionally, for large section of consumers, electricity prices have been kept below cost, partly because of consumer pressure, but partly to subsidise smaller consumers. The Electricity Act had mandated that there should be annual revision of tariff, but this did not happen for years. In 2003, in aggregate, states were charging an average billed tariff well above the cost recovery and losses were overwhelmingly driven by distribution losses. By contrast, in 2011, while the latter continued to be high, states were charging average billed tariff below cost recovery (World Bank, 2014). However, with further loans being denied to the utilities there was some realisation of the need to change. In 2012, 31 states/UTs raised tariff by an average of 16 per cent (Tamil Nadu 37 per cent, after 8 years; Kerala 30 per cent; Maharashtra 28 per cent). Eighteen states did in 2013. The short-lived government in Delhi in early 2014 made reduction in power tariffs its prime policy. This led to repercussions elsewhere in reduction of tariffs further. This will make the task of rationalising the tariff structure more difficult. State Power Regulators should take action to protect genuine consumer interest, ensuring rational tariffs and improving distributional efficiencies. But the most urgent reform seems now to develop a political consensus, both on the tariff structure at all levels and on metering every consumer and reducing both theft and genuine transmission losses. Frequent remarks that promise free or subsidised electricity are not conducive to creating the right public perception of energy as a commodity, which has a cost of production, not an entitlement. India's policy objective of inclusive development and provision of affordable energy should be achieved without sacrificing business viability. This perception is the foundation of a properly functioning energy market (Ahn and Dagmar, 2012).

Financial Situation of Utilities

The rising cost of power not followed by commensurate increase in tariff is leading to a situation of financial bankruptcy of the utilities. The 12th Five Year Plan document has shown that Distribution Companies have not been able to recover the cost of supply through tariff, and the gap between average cost of supply (ACS) and average revenue realised (ARR) has been widening slowly but surely over the years from being 76 paise in 1998–99 to 145 paise in 2009–10. It has estimated that in absence of government subsidy the commercial losses of Discoms in 2010–11 were around ₹60,000 crore (Table 3.1) (Government of

Table 3.1
Viability of major state utilities (excluding Delhi and Odisha)

	2007–08 Actual	2008–09 Actual	2009–10 Prov.	2010–11 Revised Estimates
Energy sold/energy available (%)	72.86	74.55	74.33	76.21
Revenue from sale of electricity (₹ crore)	131,220	148,605	163,475	192,827
Total cost of electricity sold (₹ crore)	174,452	212,292	235,701	261,467
Commercial losses without subsidy (₹ crore)	33,290	52,452	60,172	59,050
Average cost of supply (paise/kWh)	405.86	464.48	480.37	485.67
Average tariff (paise/kWh)	305.29	325.13	333.17	358.18
Gap between the cost of supply and tariff (paise)	100.57	139.35	147.20	127.49

Source: 12th Five Year Plan Document.

India, 2012). In a recent study, Indian Credit Ratings Agency (ICRA) has estimated losses for Discoms at ₹80,000 crore in the financial year (FY) 2012–13. The study further points out that even with government subsidy support, Discom losses will be around ₹38,000 crore (Ghosh et al., 2012). The World Bank has estimated that the power sector faces annual losses of $27 billion by 2017 unless sweeping reforms are taken to tackle inefficient subsidies and theft (Figure 3.6) (World Bank, 2014).

Sector losses have been financed by heavy borrowing. The accumulated debt of the utilities has crossed more than 2 lakh crores and increasing. This has upset the banks' capital adequacy and net worth severely. There has been increasing short-term borrowing by utilities while their credit-worthiness has been eroding. Ten years after a big bail out from the government, utilities were given another in 2012–13 several times larger than the earlier one. Even this essentially debt restructuring plan has hardly taken off two years down the line. Such

Figure 3.6
Discoms—Trends of subsidy dependence

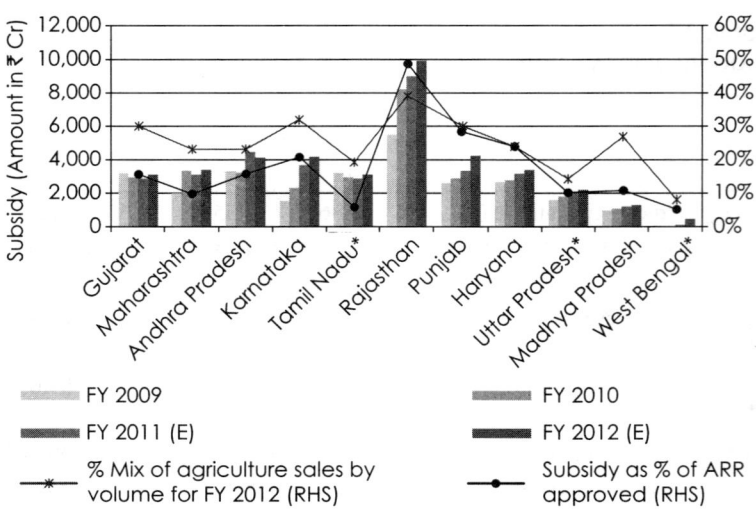

Source: ICRA Study.
*Tariff last revised in 2010.

bailouts, apart from being fiscally unsound, essentially provide limited period relief at a great cost, but the fundamental structural problems are not addressed and postponed to a never tomorrow. Further such centrally determined solutions only help state governments to delay the day of reckoning. The worrying, but perfectly reasonable, question has been raised—do we prepare a 10 lakh crore bailout package in 2020? (Chatterjee, 2013).

In this situation, the utilities are finding it increasingly difficult to procure power to meet the rising demand. As a consequence, the power deficit issue is taking a peculiar shape. In FY 2013, NTPC, with installed capacity of 41,184 MW, lost 20.4 billion units due to either low schedule or backing and 13 billion units due to fuel shortage.[34] It is indeed ironical that on the one hand there is both general and peak power shortage, and on the other, capacity already set up is not being fully utilised either because of want of fuel, and more importantly, schedule. In June 2013, average price dropped on the Indian Energy Exchange (IEX) to ₹2.027, compared with ₹4.0 a unit in the corresponding period last year. Average prices during peak hours between April and June 2013–14 fell to ₹2.8 a unit from ₹3.5 a unit a year ago though there was a 10 per cent peak deficit.[35] As such it could be inferred that 'consumer demand' is not 'power demand' because distribution companies prefer load shedding rather than buying power. No wonder Ashok Khurana, Director General, Association of Power Producers (APP) stated that instead of power off-take being function of demand, it has become a function of bankrupt Discoms.[36] This has further consequences for the power generators and the lenders.

So long as our power tariff structure will not change and consumers pay less than it costs to generate, and efficiencies will not be optimum, this situation of deficit will not ease. Recognising this, the then Deputy Chairman, Planning Commission, Montek Singh Ahluwalia, repeatedly stressed in public statements over the years that the fundamental problem of the energy sector in India was under-pricing of energy. He also made this one of the key focus areas of the 12th Plan.[37] The then Prime Minister also kept saying so in many of his speeches. Since there is enough evidence of the problems this is causing to the power sector, which is publicly recognised, it is imperative that a political consensus emerge on this critical issue.

Conclusion

What hope then for rural India? The utilities are hardly in a position to supply power to rural areas or even small urban towns. And things are not going to get much better, and not for many years. It should then be almost self-evident that the expectation to provide rural electrification through centralised grid to most of the current deficit areas is unlikely to be realised. As stated earlier, even adding more and more renewables to the grid will not be able to solve this problem satisfactorily, because of problems in supply in evening peak hours and cost of supply issues will continue to remain. If we want to increase electricity access dramatically, then alternative solutions are necessary. This is where renewables in their off-grid mode will become critical for energy access. And if we have to have renewable energy, as we must, would it not be desirable that a substantial portion of it is directly generated and supplied at the remotest point of consumption rather than generating it at large plants and losing plenty of it while travelling along long distances, and even then, not being able to supply to the last mile?

Notes

1. Planning Commission 2013. Available at http://planningcommission.nic.in/plans/planrel/12thplan/pdf/vol_2.pdf. Accessed on 22 November 2014.
2. Planning Commission 2002. Available at http://planningcommission.nic.in/plans/planrel/fiveyr/10th/volume2/v2_ch7_3.pdf. Accessed on 22 November 2014.
3. Central Electricity Authority: All India Installed Capacity (in MW) of Power Stations. Available at http://www.cea.nic.in/reports/monthly/inst_capacity/mar14.pdf Accessed on 22 November 2014.
4. BP Statistical Review of World Energy (June 2011) and IEA Key World Energy Statistics 2011.
5. IEA Key World Energy Statistics 2012.
6. As estimated in the Integrated Energy Policy Report 2006.
7. Central Electricity Authority, All India Electricity Statistics, General Review 2012.
8. Suresh Prabhu, ex-Power minister of India, who had brought in the Electricity Act 2003, 'From investor's darling, the power sector has again become a pariah' appeared in *Economic Times* on 13 August 2013.
9. Central Electricity Authority. Available at http://cea.nic.in/reports/monthly/executive_rep/installedcap_allindia.pdf. Accessed on 22 November 2014.
10. CEA. Available at http://www.cea.nic.in/reports/monthly/inst_capacity/mar14.pdf. Accessed on 22 November 2014.

11. Generation data up to 2012–13 have been taken from the All India Electricity Statistics, General Review 2013. Generation data for the power sector, excluding renewable power have been taken from CEA. Available at http://www.cea.nic.in/reports/monthly/generation_rep/tentative/mar14/opm_01.pdf. Accessed on 22 November 2014. The renewable energy share for 2013–14 has been calculated on normative basis (taking 50 per cent of the total installed capacity for 2013–14 and the capacity utilisation factors for wind, solar, biomass and small hydro as 20 per cent, 19 per cent, 50 per cent and 40 per cent, respectively).
12. Planning Commission (April 2014). The Final Report of the Expert Group on Low Carbon Strategies for Inclusive Growth. Available at http://planningcommission.nic.in/reports/genrep/rep_carbon2005.pdf. Accessed on 22 November 2014. Both the IEP and this report were chaired by the same person.
13. Ministry of Coal, Annual Report 2009–10.
14. 12th Plan Document. It is a wonder how such exact estimates are made at different times, but a definite trend is clearly visible.
15. At a KPMG seminar on 20 November 2013, L&T Infrastructure Finance stated that they have stopped lending for coal-based power projects and will be funding only renewables.
16. A dramatic picture is painted by journalistic licence of this altogether neglected issue. 'Visit North Chhattisgarh to witness the folly-strewn landscape of India's power El Dorado gone wrong. In just four districts, MoUs were signed for 64 thermal projects, amounting to approximately 50,000 MW. Of the lot, only 10,000 MW has come on line. The rest are half-built, or in some cases, limited to a signboard stuck on a vacant plot' (Jain, 2014). The problem of pollution is an invisible plot in the background!
17. CEA is blamed for not preparing/approving DPRs on time. CEA says that of the 37 DPR's approved in the last few years aggregating a capacity of 24,015 MW, in none has construction started! Clearly there are managerial issues too.
18. One unit has been commissioned in September and one is awaited.
19. Department of Atomic Energy, Annual Report 2011–12.
20. *Business Standard*, 25 July 2014.
21. The book *The Power of Promise* by M.V. Ramana has discussed in depth all issues related to development of nuclear power in India and outlined problems in developing large-scale capacity. The forecasts made have also taken from this book.
22. The World Nuclear Industry Status Report 2013. Available at http://www.worldnuclearreport.org/IMG/pdf/20130716msc-worldnuclearreport2013-lr-v4.pdf. Accessed on 22 November 2014.
23. It is interesting to note that while nuclear energy has got a 12th Plan allocation of the order of ₹60,000 crore for a capacity target of 5,500 MW, renewable energy has got around ₹19,000 crore for a target of 30,000 MW and off-grid.
24. The Power Grid Corporation of India Ltd, Gurgaon, prepared a study 'Desert Power India—2050' at the request of the MNRE in December 2013. This projected RE capacity of 485 GW by 2050. Solar power from deserts (Thar, Kutch, Ladakh, Lahaul and Spiti) alone was projected at 271 GW. The total cost of installation, transmission and balancing requirements was estimated at ₹4,374,550 crore.
25. This has led to wind potential becoming more than 100 GW compared to 49 GW earlier. There are other estimates that suggest much higher potential. Lawrence Berkley Laboratory, USA has estimated on shore wind potential in India over 800 GW!

26. The 18th Electric Power Survey of CEA suggests that electricity energy requirement will rise by 40 per cent by 2017 in 13 major cities and another 40 per cent by 2022.
27. A detailed pathway has been suggested in the Report of the Expert Group on Low Carbon Strategies (see note 12). Clearly this area is of the utmost importance for the country but is not getting adequate attention.
28. CERC tariff order dated 5 April 2013. Available at http://www.cercind.gov.in/2013/orders/SO69.pdf. Accessed on 23 November 2014.
29. The Department of Atomic Energy expects that the price from the two nuclear plants currently under discussion would be quite high when they are functional in 2021—in the range of ₹7 per unit and above.
30. Central Electricity Authority, Growth of Electricity Sector in India from 1947–2012. Available at http://www.npti.in/Download/Misc/CEA%20Growth%20of%20Electricity.pdf. Accessed on 22 November 2014. There are reports that it has now increased to 27 per cent.
31. Available at http://planningcommission.nic.in/reports/genrep/hlpf.pdf. Accessed on 22 November 2014.
32. States raise tariffs, but Discoms still in doldrums; *Business Standard*, 24 April 2013, has a detailed discussion on these issues. Available at http://www.businessstandard.com/article/economy-policy/states-raise-tariffs-but-discoms-still-in-doldrums-113042400807_1.html. Accessed on 22 November 2014.
33. As a share of state budget in 2011, this was highest in Bihar at 15 per cent—clearly a luxury which that state can ill afford (see World Bank, 2014).
34. *Business Standard*, 27 July 2013.
35. *Business Standard*, 7 June 2013: Power prices slump amid discoms' fund woes. Available at http://smartinvestor.business-standard.com/market/story-179970-storydet-Power_prices_slump_as_discom_financial_woes_transmission_contraints_rise.htm. Accessed on 22 November 2014.
36. Katya B Naidu in *Business Standard*, 27 July 2013: Power sales down as buyers black out. Available at http://smartinvestor.business-standard.com/market/story-188385-storydet-Power_sales_down_as_buyers_black_out.htm. Accessed on 22 November 2014.
37. The energy challenge in the 12th Plan is how to deal with a situation in which global energy prices will be high and the cost of alternative energy sources will also be high. Our ability to grow rapidly in this environment depends critically on our ability to transmit the high energy prices to energy users in the economy rather than keep the price artificially low (Ahluwalia, 2011).

4
Rural Electrification: Policy Landscape and Status of Access

In tandem with the approach to industrial development, electricity policies after Independence also focussed on setting up of power-generating capacity in the central and state public sector, particularly large hydro and creating a power infrastructure. The industrial sector was the focus for supply. As a result, electrified villages grew from 3,000 in 1950–51 to only 22,000 by 1960. The problem of inadequate food production and the need for self-sufficiency dominated the policy agenda in the mid-1960s. This led to a shift for power supply from rural household electrification per se to exploitation of groundwater by using electric pumps to increase agricultural production. This approach was reflected in the measure of rural electrification being extension of electricity lines to a particular area or village and not percentage of households electrified. On the one hand, this deterred household adoption of electricity, on the other, as share of agriculture in electricity increased, the financial difficulties of State Electricity Boards (SEBs) worsened as tariff was lower than costs of supply (Khandker et al., 2010). The need to address rural electricity supply in a focussed manner led to the incorporation of Rural Electrification Corporation (REC) in 1969 under the Companies Act, 1956. Its main objective was to finance and promote rural electrification projects all over the country by providing financial assistance to State Power Corporations, Electricity Departments of the state governments and rural electric co-operatives. This effort started with a focus on agriculture and irrigation. As lines were laid for providing power for the green revolution, and for connecting towns and semi-urban areas, some villages also got electricity for lighting and domestic use as a by-product. Central sector investment in the thermal generation sector started with the setting up of the National Thermal Power Corporation in 1975. The government nationalised coal mines in 1973 with the Coal Mines Nationalisation Act, and Coal India Limited (CIL) was set up to manage fuel supply to thermal power plants. With increasing problems of the environment and rehabilitation, the focus for building generation capacity shifted from hydro to thermal and nuclear power was given prominence. The latter suffered after our nuclear explosion as fuel supply was stopped.

The limited objectives for power supply did not lead to holistic development of the power sector. The low rates of growth, limited urbanisation and socialistic approach towards ownership of infrastructure and industrial assets in the public sector simply did not provide either the demand push or the environment to have greater generating capacity. There was also no serious voice raised for household supply of electricity in rural areas. Progress overall, therefore, remained tardy in all areas and the power sector gradually got into more and more trouble—of inadequate capacity; of poor generation; large transmission losses; financial deterioration of the Electricity Boards; setting up of unrealistic and below cost electricity rates, particularly in the rural agriculture sector; theft in the supply at all levels, etc. However, the turn of the century saw a commitment and many efforts to bring in reforms and the much needed thrust. Because of its importance, special attention continues to be given to the needs of this sector.

It seems that somehow, the actuality of household energy access got left behind. REC remained the main organisation to catalyse rural electricity access through creation of supply-side infrastructure, including for village household electrification. Essentially it was felt, and perhaps has since been, that financing for and building the infrastructure was the major problem. Once built, power would flow and the utilities would supply it. But the financial burdens imposed on the utilities meant that they remained unwilling partners in promoting these schemes. The next three decades saw different schemes being formulated to advance rural electrification, either directly or as part of overall rural development, though these were basically the same wine presented in different bottles. Different targets were fixed and different financial plans were prepared, but there was absence of a holistic approach and, perhaps, an understanding of the real constraints. It is no surprise that in spite of being supposedly at political centre stage, even in 2013, as evident from the recent Census 2011 results, we are still far away from our target of universal energy access. Let us visit the policy landscape over the years and try and assess both our achievements, and, honestly enough, lack of them and reasons for the same.

Policies (1974–2005)

Given below are some of the schemes that were formulated in succession.

A Minimum Needs Programme (MNP) started in the Fifth Five Year Plan (1974–79) with rural electrification as one of its components, not becoming a separate programme in itself. The SEBs were the implementing agencies. Under this programme, funds were provided as Central assistance to the states in the form of grants and loans. The areas to be covered for the purposes of rural electrification were remote, far flung and difficult villages with low load potential—precisely the ones that would be the most difficult and expensive to do with little hope of actual supply. For identifying the beneficiaries and beneficiary areas, certain all-India norms were taken as benchmarks at various points of time so that MNP assistance was directed to the population/areas with low levels of achievement below respective benchmark levels. This programme centred on low interest loans, which were attractive in the initial years. However, over a period time, the rate of interest charged increased making it uneconomical for the SEBs. This scheme continued for almost three decades without doing much and was discontinued and merged with RGGVY in 2005.

Kutir Jyoti Scheme was launched in 1988–89 with the objective of extending single-point light connections to the households of rural 'below poverty line' (BPL) families with special quotas for SC and ST families. This was the first real attempt to address household electrification, especially of the poor. The respective state governments were responsible for executing the target connections against which they received grants through REC. Under this scheme, the outlay used to be allocated amongst the states based on two indicators, i.e., size of rural population below the poverty line and level of village electrification obtained in the state in a manner that higher weightage was given to states with larger population of rural poor and with low electrification levels. A one-time cost of internal wiring and service connection charges of ₹200 per connection (which became ₹1,800 by 2002) was provided by way of 100 per cent grant to the states. However, the desired levels of success could not be achieved mainly due to unreliable and poor quality of supply as also subsequent apathy in maintenance and consequently the beneficiaries not paying for electricity bills. There were problems of estimation of electricity consumption since connections were not metered and tariff was a flat rate one, often based on connected load. There were also reports of misuse. As per REC, nearly 60 lakh households were covered under the scheme at a cost of ₹450 crore till 2004 when it was merged under the Accelerated Rural Electrification Programme (AREP) (Sreekumar and Santanu, 2011).

In 2001, the government declared the objective of 'Power for All by 2012' under Rural Electricity Supply Technology (REST) Mission. But this appears to have been more of a political announcement as no substantive action followed. The rural electrification programme became a component of the Pradhan Mantri Gramoday Yojana (PMGY) in 2001–02. This scheme was also implemented through SEBs/Electricity Departments/ Power Utilities. Funds under the programme were provided to the states as additional central assistance at 90 per cent grant and 10 per cent loan for special category states, and 30 per cent grant and 70 per cent loan for other states. The states had the discretion of utilising the funds for different components as per their own priorities. At least one *dalit*/tribal *basti* was to be included in each un-electrified village being taken up for electrification. This condition was relaxed if there were no *dalit*/tribal *basti* left to be electrified in that village. In the states where 100 per cent villages were already electrified, the implementation agencies could take up *dalit*/tribal *basti*s, and if all of them had also been electrified, they could take up electrification of hamlets or load intensification. As there was no clear-cut earmarking of percentage of funds for rural electrification, many states diverted the funds to other areas and rural electrification was neglected in the process.

Accelerated Rural Electrification Programme (AREP) brought into its ambit all the earlier schemes in 2003–04. The scope of the scheme was to provide interest subsidy of 4 per cent on loans availed by state governments/power utilities from financial institutions such as Rural Electrification Corporation (REC), Power Finance Corporation (PFC), Rural Infrastructure Development Fund (RIDF), National Agricultural Bank and Rural Development (NABARD), etc. for carrying out rural electrification programmes. This assistance was limited to electrification of un-electrified villages, electrification of hamlets/*dalit bastis*/tribal villages and electrification of households in villages through both conventional and renewable energy sources.

Accelerated electrification of 1 lakh villages and 1 crore households was launched a year later in 2004–05, merging interest subsidy scheme AREP with Kutir Jyoti Programme. For implementing this programme, District Electricity Committees were to be constituted under section 166 (5) of the Electricity Act 2003 by the state governments to facilitate proactive role for expeditious rural electrification in the district and monitor the functioning of projects. Scheme was to be implemented under overall supervision and control of REC as lead agency for the scheme.[1] The scheme had provision of providing 40 per cent capital subsidy for rural electrification projects and the balance as loan assistance on soft terms from REC. Main features of the scheme included: (a) grid-based projects as well as stand-alone projects based on distributed generation were eligible for capital subsidies; (b) capital subsidy (up to 40 per cent of capital cost) was to be linked to sustained delivery of electricity to the targeted beneficiaries over the project life of 15 years; (c) balance funds for the project were to be provided by REC as loan assistance; (d) for availing capital subsidy, projects needed to demonstrate revenue stream that resulted in sustainable operations with the given level of capital subsidy; (e) tariff was to be agreed between the beneficiaries and the Rural Electricity Supply Provider with the involvement of panchayats, cooperatives, non-governmental organisations (NGOs) and franchisees, etc; and (f) in electrified villages, 100 per cent grant was to be provided for electrification of BPL households as per existing Kutir Jyoti guidelines. The projects had a universal obligation to provide electricity to all consumers on demand and the scheme had an alignment with the policies under Sections 4 and 5 of the Electricity Act 2003 to facilitate sustainable provision of electricity in rural areas.

Electricity Act 2003

The above scheme was actually announced following the enactment of the Electricity Act 2003 (hereinafter referred as the 'Act'), which was undoubtedly the most important legislation since Independence to reform and revitalise the power sector, and was a bold attempt to address the problems in a holistic manner by providing a framework for action. This was preceded a few years earlier by the setting up of Central and State Regulatory Commissions with the intention of regulating and rationalising tariffs.[2] At the policy level, the Act under Section 6 mandated the hitherto implied Universal Service Obligation by stating that the government shall endeavour to supply electricity to all areas, including villages and hamlets.[3] Section 5 mandated the formulation of a National Policy on Rural Electrification focusing, especially, on management of local distribution networks through local institutions. Section 4 of the said Act also freed stand-alone generation, defined as an electricity system set-up to generate power and distribute electricity in a specified area without connection to the grid, and distribution networks from licensing requirements. The host of reforms and policy measures contained in the Act, viz., de-licensed generation, freedom for captive generation, recognising trading as an independent activity, introduction of open access in transmission and multiple licenses in distribution, etc., were also aimed at accelerating rural electrification. Section 2 (8) of the Act defined a captive generating plant as a power plant set-up by any person to generate electricity primarily for his own use and includes a power plant set-up by any cooperative society or association of persons for generating electricity primarily for the use of members of such cooperative society or association. This in effect is distributed generation. Section 9 (1) stipulated that a person may construct, maintain or operate a captive generating plant and dedicated transmission lines. It also provided that supply of electricity from the captive generating plant through the grid shall be regulated in the same manner as the generating station of a generating company. The Act, therefore, laid down the framework for providing rural electricity, and included non-conventional electricity as one of the sources.

Rajiv Gandhi Grameen Vidyutikaran Yojana

Meanwhile, the Central Government changed after the General Elections of 2004. The Common Minimum Programme of the United Progressive Alliance (UPA) Government, elected in 2004 on the slogan of Aam Aadmi, declared its intention to accord access to electricity a very high priority. The Congress manifesto, perhaps as a response to the accelerated electrification scheme of the previous government announced in 2004, had stated that they 'will launch a special programme so as to ensure each household will have full access to reliable power in the next three to five years'.[4] Accordingly, in order to speed up rural electrification, the Ministry of Power launched the Rajiv Gandhi Grameen Vidyutikaran Yojana (RGGVY) in March 2005. The overarching objective of the scheme was universal electricity access to rural households by 2009, a very laudable but ambitious proposition indeed. This initiative subsumed all the earlier programmes with clear provisions for capital subsidy and sufficient funds. The REC was declared as the nodal agency for implementation of the programme at the Central Government level. Apart from its role as a financial institution, it also had the prime responsibility of coordinating the rural electrification programme with the state governments, state utilities and other concerned agencies for effective implementation of schemes.

The programme components included (i) Rural Electricity Distribution Backbone (REDB); (ii) Village Electrification Infrastructure (VEI) with provision of distribution transformers of appropriate capacity in villages/habitations; and (iii) Decentralised Distributed Generation (DDG) systems (based on renewable energy sources) where grid supply is not feasible or cost-effective. It envisaged providing grid extension to 125,000 un-electrified villages, connecting about 23.4 million BPL households with a 90 per cent subsidy on connecting costs; and to augmenting the backbone network in all 462,000 electrified villages.[5] There was an expectation that while the centre built the infrastructure, the states would ensure adequate supply of power. The Office Memorandum on RGGVY stated, 'there shall be no discrimination in the hours of supply between the rural and urban households'.[6] This was a critical condition for rural electrification to succeed, but it was not honoured. Further, there was a condition that 'in the event the projects are not implemented satisfactorily in accordance with the conditionality indicated above, the capital subsidy could be converted into interest bearing loans'.[7] This also never happened. The actuality of poor supply of power was recognised when RGGVY was further renewed in February 2008 with a changed stipulation that the respective state governments shall provide guarantee for a minimum daily supply of 6–8 hours of electricity in the RGGVY network.[8] This could be largely honoured as there was no stipulation regarding which hours during the day these will be.

Soon after the announcement of RGGVY, and as per the provisions of Section 3 (1) of the Electricity Act 2003, the National Electricity Policy (NEP) was announced on 12 February 2005.[9] The Policy began by stating:

> Electricity is an essential requirement for all facets of our life. It has been recognised as a basic human need. It is a critical infrastructure on which the socio-economic development of the country depends. Supply of electricity at a reasonable rate to rural India is essential for overall development.

NEP aimed at (a) access to electricity—available for all households in next five years; (b) availability of power—demand to be fully met by 2012. Energy and peaking shortages to be overcome and adequate spinning reserve to be available; (c) supply of reliable and quality power of specified standards in an efficient manner and at reasonable rates; (d) per capita availability of electricity to be increased to over 1,000 units by 2012; (e) minimum lifeline consumption of 1 unit per household/day as a merit good by year 2012; (f) financial turnaround and commercial viability of electricity sector; and (g) protection of consumers' interests. With regard to rural electrification NEP inter-alia states that wherever grid-based electrification is not feasible (it is neither cost-effective nor the optimal solution to provide grid connectivity) decentralised distributed generation facilities together with local distribution network would be provided so that every household gets access to electricity. This would be done either through conventional or non-conventional methods of electricity generation whichever is more suitable and economical. It added that non-conventional sources of energy could be deployed even where grid connectivity exists provided it is found to be cost-effective. Further, development of infrastructure would also cater for requirement of agriculture and other economic activities, including irrigation pump sets, small and medium industries, *khadi* and village industries, cold chain and social services such as health and education. NEP further states that necessary institutional framework would need to be put in place not only to ensure creation of rural electrification infrastructure but also to operate and maintain supply system for securing reliable power supply to consumers. It suggests that responsibility of operation and maintenance and cost recovery could be discharged by utilities through appropriate arrangements with panchayats, local authorities, NGOs and other rural franchisees, etc. The provision relating to captive power plants to be set up by group of consumers was primarily aimed at enabling small and medium industries or other consumers that may not individually be in a position to set up plant of optimal size in a cost-effective manner. The Policy further states that a large number of captive and standby generating stations in India have surplus capacity that could be supplied to the grid continuously or during certain time periods. These plants offer a sizeable and potentially competitive capacity that could be harnessed for meeting demand for power. Under the Act, captive generators have access to licensees and would get access to consumers who are allowed open access. Grid inter-connections for captive generators shall be facilitated as per Section 30 of the Act.

A key outcome of the Electricity Act 2003, following the NEP, and providing policy support for the RGGVY, was the notification in 2006 of the Rural Electrification Policy.[10] It repeated its objective quite explicitly beginning by stating that 'Rural Electrification is viewed as the key for accelerating rural development. Provision of electricity is essential to cater for requirements of agriculture and other important activities including small and medium *khadi* and village industries, cold chains, health care, education, and information technology'. It envisaged grid connectivity through distribution network up to 33/11 or 66/11 kV level with appropriate development of sub-transmission and transmission systems at higher voltage levels; off-grid solutions based on stand-alone systems for villages/habitations where grid connectivity may not be feasible or cost-effective, adoption of isolated lighting technologies such as solar photovoltaic in case neither stand-alone systems nor grid connectivity are feasible. Moreover, it also stated that DDG facilities together with local distribution network based either on conventional or non-conventional methods of electricity generation whichever is more suitable and economical, could be utilised even where grid connectivity exists and electrification of BPL rural households to be financed with 100 per cent capital subsidy. It further states that the state governments should prepare and notify a Rural Electrification Plan to achieve the goal of providing access to all households. The policy made a significant change in the definition of an Electrified Village. A village would be treated as electrified if basic infrastructure was provided to the inhabited locality as well as the *dalit basti*/hamlet, electricity was provided to public places and at least 10 per cent of the total numbers of households were electrified. This was an important change, but not sufficient as we shall see later. In order to manage local distribution, the policy considered the deployment of franchisees necessary to ensure revenue sustainability and improve services to the consumers. These could be NGOs, users' associations, cooperatives or individual entrepreneurs.

Progress under RGGVY

Let us now see how the situation of electricity access has changed over the years. This would need to be seen not only for the number of villages electrified, but also households, as well as some other issues related to the quality and quantity of supply.

Under Bharat Nirman that was launched by the Government of India in 2005, a target of electrification of one lakh un-electrified villages and 1.75 crore BPL households by March, 2012 was fixed for RGGVY. Against this 1.04 lakh villages were electrified and 1.94 crore BPL households were given connection up to March 2012 (Table 4.1). As per Central Electricity Authority (CEA) Report, by the end of the Eleventh Plan, out of the total 593,732 villages in India (as per Census 2001), 556,633 villages (93.8 per cent) had been electrified (Government of India, 2012, p. 143).

The state-wise figures of village electrification are given in Table 4.2.

Needless to say, despite an overarching policy framework and a focussed programme, RGGVY could not cover all targeted villages as planned by 2009 and was further extended in 2008 into the 11th Plan period (2007–12)[12] with additional goals of supply of quality and reliable power at reasonable rates; a minimum lifeline consumption of 1 kWh per household per day as a merit good by year 2012; the deployment of franchisees to be made mandatory for rural distribution management to ensure revenue stability and DDG systems based on conventional and non-conventional energy sources where grid supply is not feasible or cost-effective. Although

Rural Electrification 53

Table 4.1
Status on RGGVY progress during 10th and 11th plans

	Un-electrified Villages (No.)			BPL Households (Lakh)		
Year	Target	Ach.	% Ach.	Target	Ach.	% Ach.
Tenth Plan						
2005–06	10,000	9,819	98.2	3	0.17	5.7
2006–07	40,000	28,706	71.8	40	6.55	16.4
Total	50,000	38,525	77.1	43	6.72	15.6
Eleventh Plan						
2007–08	10,500	9,301	88.6	40	16.21	40.5
2008–09	19,000	12,056	63.5	50	30.85	61.7
2009–10	17,500	18,374	105	47	47.18	100.4
2010–11	17,500	18,306	104.6	47	58.84	125.1
2011–12	14,500	7,934	54.7	52	34.45	66.2
Cumulative (as on 31.3.2012)	112,795	104,496	92.6	275	194.25	70.6

Source: 12th Five Year Plan Document.

Table 4.2
Village electrification status as on 31 January 2014

S. No.	States/UTs	Total Inhabited Villages as Per 2001 Census	Villages Electrified as on 31 January 2014 as Per New Definition	
			Number	Percentage (%)
1.	Andhra Pradesh	26,613	26,613	100.0
2.	Arunachal Pradesh	3,863	2,917	75.5
3.	Assam	25,124	24,156	96.1
4.	Bihar	39,015	37,854	97 0
5.	Delhi	158	158	100 0
6.	Jharkhand	29,354	26,190	89.2

(Table 4.2 Continued)

(Table 4.2 Continued)

S. No.	States/UTs	Total Inhabited Villages as Per 2001 Census	Villages Electrified as on 31 January 2014 as Per New Definition	
			Number	Percentage (%)
7.	Goa	347	347	100.0
8.	Gujarat	18,066	18,031	99.8
9.	Haryana	6,764	6,764	100.0
10.	Himachal Pradesh	17,495	17,480	99.9
11.	J&K	6,417	6,304	98.2
12.	Karnataka	27,481	27,468	99.95
13.	Kerala	1,364	1,364	100.0
14.	Madhya Pradesh	52,117	50,939	97.7
15.	Chhattisgarh	19,744	19,224	97.4
16.	Maharashtra	41,095	41,059	99.9
17.	Manipur	2,315	1,997	86.3
18.	Meghalaya	5,782	4,988	86.3
19.	Mizoram	707	661	93.5
20.	Nagaland	1,278	896	70.1
21.	Odisha	47,529	38,920	81.9
22.	Punjab	12,278	12,278	100.0
23.	Rajasthan	39,753	38,808	97.6
24.	Sikkim	450	450	100.0
25.	Tamil Nadu	15,400	15,400	100.0
26.	Tripura	858	797	92.9
27.	Uttar Pradesh	97,942	87,086	88.9
28.	Uttaranchal	15,761	15,593	98.9
29.	West Bengal	37,945	37,941	99.99
	Total (States)	**593,015**	**562,683**	**94.9**

(Table 4.2 Continued)

(Table 4.2 Continued)

S. No.	States/UTs	Total Inhabited Villages as Per 2001 Census	Villages Electrified as on 31 January 2014 as Per New Definition	
			Number	Percentage (%)
	Union Territories			
1.	Andaman & Nicobar Island	501	339	67.7
2.	Chandigarh	23	23	100.0
3.	D & N Haveli	70	70	100.0
4.	Daman & Diu	23	23	100.0
5.	Lakshadweep	8	8	100.0
6.	Pondicherry	92	92	100.0
	Total (UTs)	717	555	77.4
	Total	593,732	563,238	94.9

Source: Central Electricity Authority (2012).

a lot of work was done during this period, the scheme was later extended into the 12th Five Year Plan (2012–17) in order to ensure that the promise made in 2004 for *each and every household* to get connected is redeemed. However, universal access has still a long way to go.

The Central Electricity Authority had assessed that balance villages to be electrified as of 31 January 2014 were 30,494, of which Uttar Pradesh, Odisha, Jharkhand, Madhya Pradesh, Bihar and Chhattisgarh had the highest numbers of 10,856, 8,609, 3,164, 1,178, 1,161 and 520 villages, respectively. But it appears there is also a lot of work to be done to complete partially electrified villages.

Household Connectivity

Although most of the villages have reportedly become electrified as per the revised definition, the more serious problem of number of households electrified in terms of getting connections was left largely unaddressed as indicated by the Census 2011.

There are great disparities in electricity access among the states. For instance, while in states such as Bihar, Assam, Uttar Pradesh, Odisha and

Table 4.3
State-wise rural household electrification levels

Electrification Levels	States/Union Territories
90% and above	Himachal Pradesh (96.6%), Punjab (95.5%), Chandigarh (97.3%), NCT of Delhi (97.8%), Sikkim (90.2%), Daman & Diu (98.3%), Andhra Pradesh (89.7%), Dadra & Nagar Haveli (91.7%), Goa (95.6%), Lakshadweep (99.8%), Tamil Nadu (90.8%), Kerala (92.1%) and Puducherry (95.8%)
Between 80 and 89%	J&K (80.7%), Uttarakhand (83.1%), Haryana (87.2%), Gujarat (85%) and Karnataka (86.7%)
Between 70 and 79%	Nagaland (75.2%), Chhattisgarh (70%), Maharashtra (73.8%) and A & N Islands (79.4%)
Between 60 and 69%	Manipur (61.2%), Mizoram (68.8%) and Tripura (59.5%)
Between 50 and 59%	Rajasthan (58.3%), Meghalaya (51.6%), Arunachal Pradesh (55.5%) and Madhya Pradesh (58.3%)
Between 40 and 49%	West Bengal (40.3%)
Less than 40%	Uttar Pradesh (23.8%), Bihar (10.4%), Jharkhand (32.3%), Assam (28.4%) and Odisha (35.6%)

Source: Census of India 2011.

Jharkhand, the household electrification level was under 40 per cent, in states like Delhi, Chandigarh, Tamil Nadu, Punjab, Sikkim, Himachal Pradesh and Andhra Pradesh, the household electrification level is over 90 per cent (Table 4.3).[11]

Electricity Access Programmes and RGGVY: An Assessment

The above policies and programme were clear statements of intent and provided many elements of a frame work with a clear mandate (Patil, 2010). It would be evident that substantial infrastructure work was done in this area, especially in the last few years under the RGGVY which is clear from the number of villages electrified as per the existing definition of an electrified village. Supplies also improved in many areas. But the limitations of these achievements will become evident once one goes behind the statistics to the ground reality.

Gokak Committee Report

This was the first organised attempt to understand the pitfalls in the infrastructural approach towards electricity access.[13] This report observed that there was a downward trend for providing electricity in villages. The poor financial health of the SEBs that made them increasingly reluctant to move to rural areas because of high costs and low returns was held largely responsible for this trend. In its findings, the report mentioned that the major reasons plaguing the pace of rural electrification were: (a) very high cost of transmission lines—₹20,000–30,000 per kilometre depending on the terrain; (b) high transmission and distribution losses; (c) low and fluctuating voltage on account of the overloading of the grid system and (d) erratic power supply and poor maintenance. The report also pointed out the enormity of the financial problems posed by the subsidies in rural electrification. It quantified the four-fold increase in subsidies in the 1990s and urged efforts to contain its rising burden. As we have seen, this problem has only worsened, and considerably so. Moreover, the other problems mentioned have also not gone away, and unlikely to do so.

Major reasons attributed for low electricity access rate were summed up as follows:

i. Neglect of revenue sustainability of the additional electrification infrastructure for the rural areas made the SEBs reluctant to take up rural electrification as it led to more losses;
ii. Village electrification was left to the SEBs, which were in bad financial health and not in a position to provide sufficient funds; and
iii. The task of maintenance of rural electricity infrastructure was with the state utilities which did not have the necessary manpower in the rural areas. It concluded that substantial infrastructure became useless although this could not be quantified.

The RGGVY was designed to build upon the lessons learnt and past experience and attempted definitive action to address the pertinent issues including development of electricity access infrastructure in rural areas, increasing the viability of rural electricity infrastructure by covering all BPL families and setting up a uniform village infrastructure at block level and to cater to non-domestic demand of power, etc. However, the fundamental problems and structural weaknesses, including those pointed out by the Gokak Committee, never really got addressed.

Evaluation Reports of RGGVY 2011–12

In order to evaluate the RGGVY programme, and with a view to assess the impact and the outcome on the socio-economic fabric of rural development across the nation, the Rural Electrification Corporation commissioned studies to four agencies in 2011 covering 1,000 villages spreading across 150 districts in 20 states.[14] Major findings of the assessment studies were the following:

i. All states supply minimum 6–8 hours of electricity except Bihar, Jharkhand and some places of Rajasthan and Uttar Pradesh. Uttar Pradesh and Assam may have between 6 and 12 hours but of poor quality. Haryana has between 9 and 14 hours and AP, Maharashtra and Rajasthan between 14 and 16 hours.

ii. One evaluation found nil supply in Bihar during peak hours, while the other saw an average of about an hour. Even Maharashtra was found to have 1 hour and Haryana 2. Similar problems are there in Assam, Jharkhand and Uttar Pradesh. Supply of electricity during peak hours is generally good in Andhra Pradesh, Arunachal Pradesh, Chhattisgarh, Gujarat, Himachal Pradesh, Karnataka, Tripura and Tamil Nadu.

iii. Access of electricity was provided to almost all (96 per cent) public places, i.e., schools, panchayats, community health centres, etc. But their actual energisation is a serious issue in many states—Uttar Pradesh, Bihar, Odisha, Jharkhand and Assam.

iv. The position regarding connections to above poverty line (APL) in Uttar Pradesh, Bihar and Jharkhand is very poor. In Bihar it is 2.4 per cent, Assam 3.4 per cent and Jharkhand 12.1 per cent. The Integrated Research and Action for Development (IRADe) Report said that only 2 out of 1,536 APL households they contacted in Uttar Pradesh had connections. It also found a large number of unauthorised connections in Uttar Pradesh, Assam and Rajasthan. A large percentage of BPL households have got connections but there is still a significant gap because either all connections were not provided, or the DPRs did not include all or population growth has created more households.

v. Commercial activities observed in West Bengal, Tripura and Odisha (Tripura: bamboo mat, juice making shops, private tuitions; Rajasthan: weaving, tailoring, running *kirana* shops; Odisha: Shops). These are really insignificant.

vi. Socio-economic Impact: Children education, ease in household chores, women empowerment, sense of security and comfort has improved. (Actually, these are primarily benefits of lighting, and can come from solar energy too.)

vii. 11 kV system is sufficient for all domestic loads.

viii. DT capacity is adequate for all sanctioned numbers of BPL consumers (with an approved single point connected load of 40/60 W) and a few of APL consumers. In general, however, except for some states, it is not sufficient for proper load. Transformer burning happens frequently.

ix. Franchisees are very few and basically collecting revenue. There is a perceived skill shortage.

Greenpeace had done a Social Survey Report in one district each of Uttar Pradesh (Azamgarh)[15] and Andhra Pradesh (Srikakulam)[16] and two districts of Bihar (Saran and Madhubani),[17] the results of which were published in April 2011. Given below are their brief observations:

Andhra Pradesh: Most villages were electrified during 1976–91. Electrification work is largely completed. Villagers complained that the electricity supply was erratic and available at times such as afternoon when they had no real use for it. It is quite unreliable in terms of quality and low voltage. It comes for about 8 hours a day and they have supply for 20–25 days a month. Households were paying in the range of ₹100–300 every month and also all households were using and spending more than ₹50 every month on kerosene as an alternative source of lighting.

Uttar Pradesh: Only 48 per cent (50,191 out of 104,603) BPL households entitled for free connection received this facility. They are getting supply for 6–8 hours available only in late evening when it is not required.

APL families had given ₹5,000–12,000 to get a connection. Most people are getting electricity through illegal connections; the practice of which is rampant in that area. Infrastructure in villages has been developed only to meet 10 per cent of the target. Low voltage and frequent burn out and non-functionality of transformers have further compounded the problem. Total 97 per cent households covered under RGGVY were using and spending on kerosene as an alternative source of lighting. Total 96 per cent revealed that they had not taken electricity connection for irrigation. Beneficiaries are now receiving bills with cumulative charges, which have caused much unrest.

Saran, Bihar: A large number of families belonging to BPL category had not received electricity connection. Total 87 per cent people surveyed complained of erratic supply, often available only at midnight when they had no use for it. It is highly unreliable and of low voltage, which rendered the connections useless for them. In all, 97.9 per cent of the households covered by RGGVY were using and spending on kerosene for lighting, 50 per cent spending ₹50–100 per month and 28 per cent between ₹100–300. The newly laid transformers had burnt soon after installation. There is a high incidence of theft. None of the villages were receiving 6–8 hours of electricity and supply was restricted to less than 10 days in a month. Irrigation is rain-fed or through diesel generators.

Madhubani, Bihar: In all, 68.4 per cent BPL listed had received connection; 95.3 per cent of these reported electricity supply to be erratic, unreliable and of low voltage, often available after midnight when they had no use for it. Total 87.5 per cent of households covered under RGGVY were using kerosene and spending up to ₹100. New transformers burnt soon after installation. Irrigation was rain-fed or through diesel generators (36 per cent).

Issues in Rural Electrification

The observations made above and the findings reflect the ground position as late as 2011–12. These findings are corroborated by many smaller studies/surveys, field visits and anecdotal evidence. These touch upon the heart of the problem and also identify the implementation challenges.

It is only when these are recognised and understood that the real problems can be addressed and possible solutions found. Therefore, they need to be discussed in some detail.

Hours of supply: The biggest problem is the hours of supply, both in terms of the total number, and more importantly, at which time during the day and night. RGGVY envisaged 6–8 hours of supply, but never specified which hours these should be during the day/night. As already noted, several states are unable to supply for even these number of hours. But, in many other states, where the supply is for more number of hours, there is a general anecdotal understanding that most of these hours are actually at night and, in an unreliable way, power keeps coming and going during the day. These are times when the households really do not need it. The evaluations bear this out quite clearly. Unfortunately, there is no detailed analysis of what are the actual hours of supply and what percentage of the areas get what hours. Nevertheless, there is little doubt that the fundamental issue of concern is that power is usually not forthcoming during the evening hours, i.e., 6–10 p.m. when there is lighting requirement. Howsoever much we may complete transmission and distribution systems and claim electrification coverage of villages, this concern of no electricity, particularly during evening hours, in no way looks like going away in major parts of the country, certainly not for problem regions, because of power shortages. That is why the issue of future power demand and supply in the country was analysed at length in the previous chapter. The only way this can be done is by compulsorily feeding power to rural areas during peak hours, which simply does not seem to be possible.

Electrification of hamlets: Being hidden away is the problem of hamlets. Initially, those hamlets with a population of more than 300 were supposed to be covered by the grid supply. In 2008, this limit was reduced to 100. Progress in hamlets between 100 and 300 population has been slow and work on smaller hamlets has not begun, except in states that had a separate programme. Since each and every household has to be reached, these habitations are now proposed to be taken up by REC in the 12th Plan, although there is no clarity on what the numbers are or what should be the norms regarding the distance from the grid or the minimum number of households. But it appears that covering hamlets is only going to compound the problem—much greater costs of transmission and distribution infrastructure, and of supply, without the commensurate benefits—all the problems, and more, of supplying electricity to the villages. Incidentally, and perhaps naturally, most of these hamlets are also in the states of Uttar Pradesh, Bihar, Odisha, Assam, Jharkhand and the north eastern (NE) states.

Household electrification: It has been noted earlier that while most villages may have been electrified, it does not necessarily mean that most households have also been electrified. In 2011, with 91 per cent villages electrified, 68 per cent of the target of BPL connections was met but overall household electrification went up only from 43 per cent to 56 per cent (Sreekumar and Santanu, 2011, p. 1). The problem of BPL household access has been partly resolved technically by providing free connections, although the problem of providing them electricity remains, because nobody really knows how many get how much power, or whether they pay for what they get. Under the RGGVY, 16.47 million BPL households were given connections.[18] The real problem lies with APL connections, as confirmed by the findings of the evaluations. These connections are voluntary. These obviously have been slow where supply of power is poor, and APL houses do not perceive much benefit especially because one has to pay for the connection and the subsequent monthly fixed charges irrespective of the actual supply. APL households have not been approaching the distribution companies for connections. Only 15 lakh households were connected in the previous 6 years, which is only 3 per cent of the total APL target of 5.46 crore (Sreekumar and Santanu, 2011, p. 11). Villages without power outages have an electrification rate of 81 per cent. In the Indian rural electricity scenario, household connection is the key issue. For this reliability and sustainability of electricity, supply is the key (Khandker et al., 2010). How will such households be covered in the future?

It is also important to note that looking at progress in electricity access in percentage terms could be very deceptive when the base is getting larger. In 2011, around 74 million rural households were still without access to modern lighting services (Figure 4.1), which is only a marginal decline from about 78 million in 2001. Between censuses 2001 and 2011, total number of rural households increased by about 29.5 million while household connectivity increased by about 32.5 million. Therefore, there has not been much change in absolute numbers. And it is this number that may not also substantially change in the near future if

Figure 4.1
Households by main source of lighting

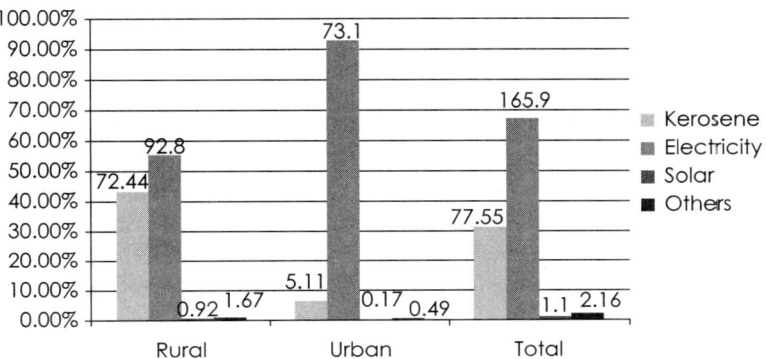

Source: Census of India 2011.
Note: Figures given on top of the bars are absolute numbers in millions.

we continue with business as usual. This is partly because rural population growth is also likely to be the highest in areas where the problem of electricity connections and supply is also the most.

Geographical dispersal: It is clear from Table 4.3 that the problem of energy access is not universal in nature in the country in terms of spread and there are pockets, which are extremely good like Gujarat and Himachal Pradesh. The crux of the problem lies in Bihar, followed by Uttar Pradesh, Jharkhand and parts of Assam, Odisha, Madhya Pradesh and the north east states. There would also be pockets or habitations[19] or households spread across geographical areas, in other states. It is all these places or areas that are unlikely to be serviced by the centralised grid in the near future. Therefore, when we think of solutions to the problem of energy access and look at alternative means of supplying power, we will have to focus on these areas. Our approach will necessarily have to be disaggregated.

Transformer capacity: The lack of transformer capacity has also been a big area of concern and is actually reflective of the symbolic nature of village electrification. Transformers have been sized taking BPL household loads at 40/60 W. Actually, they should have been sized keeping in mind the possible APL load also, if there was a realistic expectation that there would be sufficient supply of electricity, and that APL connections would actually be taken. Evaluation reports have confirmed that, in certain pockets, where connections are low, transformer outages have been very high. It would be useful to draw a correlation with theft of power when supplies actually come studying drawing of power beyond the installed transformer capacity. And we know anecdotally, when a transformer gets burnt, then for many days there may be no power as transformer replacements in such areas are not the easiest things to do and take their own time. Bihar Chief Minister, Nitish Kumar has been quoted stating that 'the low-capacity power transformers (16 and 25 kVA) installed in villages will not serve any purpose. With these transformers, we cannot use electricity even for irrigation. Thousands of such transformers have been burnt' (De, 2013). The fact is established, but not the reasons for it. This will continue to be a major problem.

This leads to a larger issue. The fact is that larger-sized transformers should have been used to cater to three phase loads and agricultural loads. It emphasises the fact that, in a large number of places, electrification has not led to the kind of localised socio-economic development that was expected, whether it is micro-industries or irrigation. This is also admitted in the 12th Plan document. This is because power is not enough, or for sufficient number of hours, and that too continuously, or reliable or of proper quality in many areas. This problem is also, therefore, likely to remain. The Planning Commission also recognised the focus of RGGVY on household connectivity without addressing the larger problem and needs of electricity supply. The 12th Plan intends to resolve this by restructuring the RGGVY, but how will the rural network be strengthened is the question. The RGGVY thus has not been able to internalise the comprehensive idea of energy security for the poor described in the IEP 2006 which essentially means poor should be provided energy beyond subsistence needs so that livelihood opportunities can be enhanced (Sreekumar and Santanu, 2011, p. 18). Shri Ajay Shankar, who served the Ministry of Power at a very senior position and at one point of time was the nodal officer in the Ministry for RGGVY, had observed that while urban India engages productively with the global economy, in the absence of electricity, economic activity in our villages has to be necessarily confined to the pre-industrial era (Shankar, 2013).

Rural Electrification 65

Public facilities: Although the evaluation says 96 per cent of the public facilities had access to electricity, the state-wise observations suggest that in many states the connections are not taken. Since many governmental agencies also have to get the connections and pay the bills, the absence of these in many states suggests that there is really little confidence that there will be regular supply. It also creates a poor environment where health and educational facilities are without power and where Block Offices and Police Stations and Banks have to depend upon diesel generators or suffer substantive power cuts. This problem also needs to be addressed and can be done quite adequately with solar roof top for offices during the day and lighting by home light systems for the residences in the evening.

Definition of village electrification: This leads to the heart of the statistical problem. The definition of village electrification has kept on changing over time. Initially a village was deemed to be electrified if electricity was used within its revenue area for any purpose whatsoever. This was in line with the thrust on providing irrigation through electrification of pumps. In October 1997, the definition was changed to bring back the focus on household electrification. A village was deemed to be electrified if electricity was used in the inhabited locality within the revenue boundary of the village for any purpose whatsoever. A new and more precise definition of village electrification was finalised in April 2004.[20] As per the new definition, a village would be declared as electrified if: (a) basic infrastructure such as distribution transformer and distribution lines are provided in the habited locality as well as the *dalit basti/* hamlet where it exists; (b) electricity is provided to public places such as schools, panchayat offices, health centres, dispensaries, community centres, etc.; (c) the number of households electrified should be at least 10 per cent of the total number of households in the village; and (d) the panchayat should certify that the village is electrified. Although the new definition was more encompassing as also target-specific, it still did not capture the core of the problem, which related to actual supplies and a more demand-based approach.

It is time for a reality check. Real rural electrification levels in the country would be better estimated if the definition would be more specific

- More than 50 per cent of all households and public places have actual connection;

- Electricity is supplied for a minimum of 12 hours, of which there is a minimum of 3 hours during 5.30–10.30 p.m.;
- The transformer size is related to the potential load of more than 75 per cent of the population based on the stated objective of providing 1 kWh per household per day; and
- Billing to and collection should be from more than 80 per cent of the connections.

Such a definition would immediately give a very different picture of rural electrification in India and the number of villages deemed electrified may drop substantially. Perhaps this would force us to look at alternative solutions which go beyond the objective of simple provision of infrastructure, inadequate as it might be, and go to actual provision of services. The Parliamentary Standing Committee on Energy observed that

> ... the real objective ... (of the scheme) ... is to provide electricity and illuminate the households rather than to just erect poles and install transformers.... There is no justification for incurring huge sums on the infrastructure alone when it cannot be taken to its logical end i.e. supply of electricity to (all) households.[21]

An honest exercise to find electrified villages based on this definition has naturally not been done, but it would be well advised.

Critique of approach: What went wrong? The focus of energy planning has always been on the 'supply side' with emphasis on 'hardware' aspects and large projects. The demand side aspects of energy, which is also the key to energy access, have generally been neglected. One could say that household electrification was considered as a by-product of the conventional electricity development plans based on commercial considerations, and universal electrification of all villages and all households was expected to be achieved in some distant future as a result of the trickle-down effect. No holistic policy for rural electrification emerged. Although this area got somewhat focussed attention in the last decade, the approach did not change much (Patil, 2010).

At the outset, rural electrification did not have an exclusive institutional structure. The instruments for implementation remained the SEBs for whom rural electrification was not a priority. They were essentially engaged in the business of electricity supply. The state utilities in general were happy to build the transmission infrastructure, but were not very interested, and also did not have the competency, to undertake this task in a holistic manner. The Central Agency was concerned only with giving funds to provide infrastructure and was never concerned with the question of service delivery and the sufficiency of power, which was left in the hands of the states.

The approach also suggested an assumption that only sufficient finance and building the infrastructure was needed. Neither the centre nor many states looked into the question of adequacy of supply, of its quality and of its cost. Often in India we go for the symbolic, rather than the real. Energy access infrastructure in many parts of the country is a good example. It was perhaps felt that energy access would be achieved by providing infrastructure and giving BPL households connections. The distribution infrastructure was actually designed to meet *only* the load of the BPL connections given, not to meet the possible loads, which would result from proper supply. Supply of electricity was never adequate, nor the system capable of fully supplying it, or even distributing it. So supply became partial and unreliable. The definition of an electrified village based on connections served to mask this reality.

The electric utilities are obviously not keen to energise villages or supply power to them. This has been discussed earlier. If the actual cost of supply of power is so high and the tariff set is minimal, then the more you supply the more you lose. What then was the incentive for supply of power to rural areas? The utilities have already accumulated losses of thousands of crores.[22] How can they continue to lose in this manner? The utilities would want to supply all the power they have to the urban areas first where they would get a reasonable tariff. In a situation of low supplies, the power will go to the areas where payment is highest. Strategically, this is the correct approach.

So utilities delay connections in rural areas and applications, such as the ones which come, keep getting delayed, which are cited as one reason to have poor APL connections. But there is a general neglect too—of billing, of collection, of maintenance, etc., which makes the situation only worse—in a sense the rural areas have been left to themselves, both in terms of supply and collections, and more so in areas where the problems are more acute. As time goes on, and the cost of procuring power goes up, rural energy supply through the utilities will only become that much more difficult.

Since supply is not regular or reliable or enough, people either do not take connections or do not pay for the supply or steal whenever the power does come. This only further vitiates the atmosphere making future sustainable supplies, which include reasonable tariffs and their regular payment, that much more difficult and challenging.

There is another problem too. Without discussing the ongoing debate on the poverty line in respect of income expenditure and the percentage of people below it, we refer to the poverty link basket (PLB) where a person's consumption expenditure being less than ₹578 per month includes an expenditure of ₹70 per month on fuel.[23] Around 42 per cent of the rural population was estimated to be below this poverty line at that time. For poorer populations, having electricity connections, and paying connection charges, and, more importantly, electricity bills along with fixed charges for erratic power supply, are considered as key barriers. These barriers will not go away. This is one reason why states with higher GDP have a greater penetration level of the grid. The states of Bihar, Jharkhand, Odisha, Uttar Pradesh and Assam, which have a low per capita GDP, also have low levels of household electrification. This may also explain partly the high level of unauthorised drawing of power in these areas.

Franchisee arrangements: Under RGGVY, franchisee-based model was a key component for ensuring revenue sustainability, and management of rural distribution networks. Franchisee may be one of the defined entities empowered by the state to develop/operate a generation and distribution system. This may involve the creation of local infrastructure and institutions to support electricity distribution in an identified contiguous area for a prescribed duration and collect revenues directly from consumers. The franchisee would be responsible for repair and maintenance, new connections, billing and collection, monitoring and reporting. Different franchisee models are given in the franchisee guidelines. However, the franchisee model simply has not taken off. Even though the RGGVY scheme prioritised on a model that was expected to promote 'entrepreneurship' among the franchisees, the implemented model became a simplistic 'revenue collection' franchisee. There are some very successful examples, one such being Assam. It may be useful to briefly review its working. There is also a successful model in Bhiwandi and Torrent Power has taken up this challenging task in the city of Agra in Uttar Pradesh.

Rural Electrification 69

Box 4.1 The franchisee system in Assam

The franchisees in Assam are individual entrepreneurs. Under the Single Point Power Supply (SPPS) scheme, measured electrical energy at the LT side of distribution transformers (DTR) is given to franchisee (Agents), for revenue collection in the command areas of DTR.

The SPPS scheme is practised throughout Assam as a franchisee model for villages. In the present form, out of the 100 units at DTR outlet, 90 units are expected to be billed and 15 per cent of the total billed amount, i.e., 13.5 units, are given as a commission to the franchisee.

The field survey of sample villages in Bongaigaon Distribution Circle suggests that average billing out of the total units supplied is about 87–90 per cent, which is a very good showing. Franchisees in the rural areas have definitely shown very good results in terms of improvement in revenue collections. Among the sample villages it was seen that almost 100 per cent of the villages were fully electrified with the electrification certificate issued by the Gram Panchayat and access to electricity was available to 34 per cent of the households who were having domestic connections. If shortage of power supply is overcome, further intensification of consumers is feasible.

Franchisees employ local persons. They remarkably improved billing and revenue collections. They are willing to expand their operations. Further planning by franchisee to cater to increasing demand by strengthening the network is impracticable network. Considerable improvements are noticed in attending to faults and repairs in the network, and time of reporting faults as well as attending to faults has been substantially reduced.

Source: Evaluation of Franchisee System; Report on Bongaigaon, Assam (2007): conducted by IRADe for Rural Electrification Corporation. Available at http://www.recindia.nic.in/download/Franchisee_Eval_Bongaigaon.pdf. Accessed on 22 November 2014.

It would work where revenue collection has been poor but electricity is being supplied. And indeed that can and should be encouraged for the reason that revenue collection should be increased in such areas. However, the model becomes fundamentally flawed when additional electricity supply gets involved. The vicious cycle leading to the utility not wanting to supply to rural areas is because of the difference between the cost of supply and the revenues realised. As long as they do not match, supplies will always be constrained. A franchisee can reduce the cost of collection and maintenance, and earn money in the process, but the cushion is generally not there for the utility. And also satisfactory supply of power is not in the hands of the franchisee. While targets for establishing franchisees have been generally met, a sustainable revenue model has not yet been established anywhere. There have been evaluations by TERI and IRADe in 2007 and 2011, but they have examined only the actual results not the alternatives and possibilities. It has been suggested that a broader mandate of livelihood facilitation be given to the franchisees, but it appears that the structural difficulties of rural electrification will remain too big a barrier.

The above analysis raises several questions. First, whether the full infrastructure of the centralised grid to achieve universal access will be built, or whether it can, or even, should. Second, should there be an analysis of the total costs already incurred and further requirement of resources simply to build the transmission infrastructure, and more importantly, the distribution infrastructure required to cater to a realistic expected load? Third, should there be an analysis of where the infrastructure built has become infructuous, and how would that be restored? It is essential that these questions be addressed before we go to the last mile, and especially for the areas where the problems are the greatest. With this scenario, it is almost impossible not to accept that we need to seriously consider alternative approaches for electricity access which we strongly believe would essentially come from renewable energy sources. In this context, these sources can no longer be seen as merely complementary to centralised grid energy supply, but possibly the only long term way to rural electrification, at least for certain identified large areas within the country. Clearly an integrated and carefully calibrated approach would be needed rather than a one-size-fits-all approach. We, therefore, now turn to see what has happened in the past, the policy framework and possibilities and constraints of the use of renewable energy for electricity access.

Notes

1. In India, steps for the formation of rural electric co-operatives for the distribution of power in rural areas for the first time were taken up in the latter half of the 1960s when the Government of India sponsored an investigation by an expert team from the NRECA (National Rural Electric Co-operatives Association), USA, for identifying a few areas with adequate potential for the establishment of rural electric cooperatives. However, there was no tangible process in this direction.

Rural Electrification   71

2. In this area, they have not been able to do much, though of late there is some movement but tariff rationalization continues to be a major problem because of its politicisation.
3. The Gazette of India, Extraordinary, No 39, 2 June 2003: Central Electricity Authority. Available at http://www.cea.nic.in/reports/electricity_act2003.pdf. Accessed on 22 November 2014.
4. Congress Manifesto 2004. Available at http://www.congresssandesh.com/manifesto-2004/16.html. Accessed on 22 November 2014.
5. Rajiv Gandhi Grameen Vidyutikaran Yojana Scheme of Rural Electricity Infrastructure and Household Electrification, Ministry of Power, Office Memorandum No. 44/19/2004-D(RE) dated 18 March 2005. Available at http://rggvy.gov.in/rggvy/rggvyportal/office_memo_mar05.htm. Accessed on 22 November 2014.
6. Ibid.
7. Ibid. Further, there was a condition that 'in the event the projects are not implemented satisfactorily in accordance with the conditionality indicated above, the capital subsidy could be converted into interest bearing loans'. This never happened, nor was the condition met.
8. Ministry of Power's OM No. File No.44/37/07-D(RE) dated 6 February 2008. Available at http://rggvy.gov.in/rggvy/rggvyportal/office_memo_feb08.htm. Accessed on 22 November 2014.
9. Ministry of Power's No. 23/40/2004-R&R (Vol. II) dated 12 February 2005. Available at http://www.powermin.nic.in/whats_new/national_electricity_policy.htm. Accessed on 22 November 2014.
10. The Gazette of India, Extraordinary, Part-I-Section-1, No 44/26/05-RE (Vol II) dated 23 August 2006, Ministry of Power, Government of India.
11. Although the Census 2011 data indicates that only around 55.3 per cent rural households were having access to electricity, the NSS 66th Round Survey conducted by National Sample Survey Organisation (NSSO) for 2009–10 shows that access to electricity in urban households was 94 per cent and in rural households was 67.3 per cent. However, the quality of electricity supply continues to remain an area of concern, particularly in rural areas, where the consumers get supplies for less than 8 hours a day in certain states. Though 67 per cent of the rural households were reported to have access to electricity in 2009–10, their per capita consumption of electricity is only around 8 units per month, which is just one third of reported consumption of 24 units in urban areas. This is because of poor quality of electricity supplies.
12. Ministry of Power, Order No. 44/37/07-D(RE) dated 6 February 2008.
13. Gokak Committee Report on Distribued Generation. Ministry of Power, Government of India. 2003. Available at http://www.powermin.nic.in/reports/pdf/gokak_report.pdf. Accessed on 22 November 2014.
14. The Rural Electrification Corporation (REC) hired the services of TERI, IRADe, Sambodhi Research and Communications and Tetratech Ltd for evaluation of RGGVY.
15. Available at http://www.greenpeace.org/india/Global/india/report/RGGVY-_Azamgarh_Report_FINAL.pdf. Accessed on 22 November 2014.
16. Available at http://www.greenpeace.org/india/Global/india/report/AP%20RGGVY%20social%20audit%20report%20FINAL.pdf. Accessed on 22 November 2014.
17. Available at http://www.greenpeace.org/india/Global/india/RGGVY_Bihar-Saran_survey_report%20FINAL.pdf and http://www.greenpeace.org/india/Global/india/report/RGGVY-_Madhubani_Report%20FINAL.pdf. Accessed on 22 November 2014.

18. REC figures quoted in 'Going Remote RR inventing the off-grid solar revolution for clean energy for all'. CSE Delhi 2012.
19. Habitations with population below one hundred are not included as per the approved RGGVY programme. As per REC it is proposed to cover the left out habitations, if any, in the phase II of the programme, whose commencement is yet to be decided by the Government of India (Source: Ministry of Power: http://rggvy.gov.in/rggvy/rggvyportal/rggvy_ques.html, accessed on 22 November 2014).
20. Ministry of Power's OM No. 42/1/2001-D(RE) dated 5 February 2004.
21. Standing Committee on Energy—41st Report, Lok Sabha Secretariat, December 2013.
22. The latest figures have been mentioned in Chapter 3.
23. Planning Commission (2009): Report of the Expert Group to Review the Methodology for Estimation of Poverty.

5
Renewable Energy Options for Electricity Access

Electricity access can be achieved in three ways—grid extension (done under the RGGVY scheme), which could include franchisee models and tail end projects feeding into the grid; small-scale DDG (used by local entrepreneurs for diesel-based electricity generation and also renewable energy–based systems—though these are currently niche applications, these are the likely models for the future); and household-level technology (solar PV-based home lighting systems that primarily meet lighting energy needs and also other low consumption activities, with minimal unit investment costs and organisational challenges). The policy landscape for electricity access has so far been grid-centric and supply oriented. There has not been any real policy that brings renewable energy–based electricity access to the centre stage. Renewable energy technologies have always been considered a secondary option, or an interim solution. The larger programmes of renewable energy that have been tried have been largely government-led, although there are now many small examples and models led by private initiative. Deployment levels have been linked with annual budgetary resources or some form of funding from bilateral agencies or other sources, both of which have been less than sufficient. A lot of experimentation has been done and much has been achieved in recent years. Perhaps the Ministry of New and Renewable Energy (MNRE) has not got sufficient credit for this work. It has of course been argued that 'all experiments to provide clean energy to the energy-deprived through a non-grid based model remain exactly that: largely experimental, (Centre for Science and Environment, 2012). However, it is felt that enough information is now available. Let us briefly see the various kinds of efforts made and different models in India to find possible renewable energy solutions and policy requirements for electricity access, which will be sustainable and can be upscaled. We first discuss the home lighting systems, then the mini-grid and later grid-based models. But before that, let us briefly look at the global picture.

Pakistan was one of the first countries to experiment with solar mini-grids setting them up in eight villages in the late 1980s. They failed because it was managed by the community and was not cost-effective. By the mid-1990s, enough entrepreneurs and aid agencies had experimented with solar for people to realise its potential promises. But its diffusion continued to be slow. The GE facility and the World Bank estimated that by 2000, only 500,000 solar systems were in use and almost 0.6 million solar lanterns (Martinet et al., 2001). But the next few years saw acceleration. In Sri Lanka between 1982 and 1999, only 6,000 systems were sold, but between 2000 and 2006 this number reached 1 lakh. Similar surges took place in India, Indonesia, Bangladesh and China. This effort was not led by large corporations, but by early pioneers, who were independent individuals driven by a compelling vision setting out to effect profound technological change. Though they achieved success, their efforts and achievements were restricted by difficulties in mobilising capital to invest in market infrastructure and for consumer finance to make solar affordable to their customers (Miller, 2009).

The Remote Village Electrification Programme

Coverage

The largest programme has been the Village Electrification Programme started by the Ministry of New and Renewable Energy in 2001 for provision of basic lighting facilities in un-electrified census villages and hamlets by 2012.[1] With the announcement of the RGGVY in March 2005 the scheme was re-aligned to only cover those villages/habitations where grid connectivity would not be feasible or cost-effective, although in a sense, feasibility and cost-effectiveness have never been clearly defined. Basically, therefore, the objective was to provide electricity through renewable energy sources to those un-electrified remote census villages and remote un-electrified hamlets of electrified census villages not covered under RGGVY. The programme envisaged covering all households in such villages/hamlets and creation of capability for accessing a minimum of 1 kWh/household/day. Under the RVE programme, solar photovoltaic (SPV) home lighting systems, small hydropower plants, biomass gasification systems in conjunction with 100 per cent producer gas engines or with dual-fuel engines using non-edible vegetable oils, non-edible vegetable oil-based engines, biogas engines, SPV power plants were to be deployed. However, most of the electrification under the programme has been provided only with SPV home lighting systems, because of which many people have often called RVE largely a solar lighting programme.[2] Initially, two 9–11 W CFL lights given to each household from 2010 onwards LED lights were also permitted. However, although there is a natural desire to get electricity through grid, and RVE has been generally considered as an interim solution, the reality is that for the households in concerned RVE villages lighting from solar became available in the evenings when it was most needed. This was better than the power available in thousands of so-called electrified villages who continued to mostly use kerosene.

This programme was funded by the MNRE but implemented by the states. Villages were selected based on proposals sent by the state governments. REC vetted the list of villages sent and only those villages were selected, which were not part of the DPRs prepared under RGGVY. The states chose an implementation agency, mostly the State Renewable Agencies. Projects were bid out. Implementation was carried out by the selected technical partner and the agency. In all, 90 per cent of the funds were met by the Government of India. The balance 10 per cent was met by the users or the state or shared. The state government had to give a commitment to replace the batteries after a few years.

The progress of sanction and completion of villages up to the 11th Plan and year wise during the 11th Plan (2007–12) are given at Table 5.1.

Table 5.1
Year-wise state-wise remote village electrification

S. No.	State	Villages								
		Up to 2006–07		During 11th Plan		2012–13		2013–14		Cumulative	
		S	C	S	C	S	C	S	C	S	C
1.	Arunachal Pradesh	297	156		141					297	297
2.	Assam	501	39	1,691	1,817		27		39	2,192	1,922
3.	Chhattisgarh	368	325	314	243					682	568
4.	Gujarat	38	2		36					38	38
5.	Himachal Pradesh	21	1		20					21	21
6.	J&K	140	117	300	43	11	174	25		476	334
7.	Jharkhand	469	287	251	206					720	493
8.	Karnataka	20		2	16					22	16
9.	Madhya Pradesh	199	30	424	351		134		62	623	577
10.	Maharashtra	271	110	82	228		2			353	340
11.	Manipur	191	134	46	103					237	237
12.	Meghalaya	97	25	66	124					163	149
13.	Mizoram	20	20							20	20
14.	Nagaland	3		8	11					11	11

S. No.	State	Up to 2006–07		During 11th Plan		2012–13		2013–14		Cumulative	
		S	C	S	C	S	C	S	C	S	C
15.	Odisha	215	18	1,505	584				549	1,720	1,495
16.	Rajasthan	327	129	13	163					340	292
17.	Tripura	62	58	23	2					85	60
18.	Uttarakhand	593	396	78	76					671	476
19.	Uttar Pradesh	250	0	34	98				15	284	113
20.	West Bengal	1,177	1,171	24	6					1,201	1,177
	Total	5,259	3,018	4,861	4,268	11	685	25	665	10,156	8,636

Hamlets

S. No.	State	Up to 2006–07		During 11th Plan		2012–13		2013–14		Cumulative	
		S	C	S	C	S	C	S	C	S	C
1.	Andhra Pradesh			13	13					13	13
2.	Arunachal Pradesh	1								1	0
3.	Goa			19				19		19	19
4.	Haryana	194	45	92	241					286	286
5.	Himachal Pradesh	1								1	0
6.	J&K			20			15			20	15
7.	Karnataka			57	14					57	14
8.	Kerala	558	558	19	49					607	607

(Table 5.1 Continued)

(Table 5.1 Continued)

		Hamlets								Cumulative	
		Up to 2006–07		During 11th Plan		2012–13		2013–14			
S. No.	State	S	C	S	C	S	C	S	C	S	C
9.	Manipur			3	3					3	3
10.	Odisha			23			4			23	4
11.	Rajasthan			90			90			90	90
12.	Sikkim	13	13							13	13
13.	Tamil Nadu	152	101	32			30			184	131
14.	Tripura	488	178	456	537		67			944	782
15.	Uttarakhand	52	34	95			84			147	118
16.	Uttar Pradesh			223	86					223	86
17.	West Bengal	9	2							9	2
	Total	1,468	931	1,172	943	0	290	0	19	2,640	2,183

Source: Ministry of New and Renewable Energy. S = sanctioned; C = Completed.

Renewable Energy Options for Electricity Access 79

By March 2012, the cumulative sanction under the RVE programme had thus reached around 10,000 villages and hamlets[3] with the remaining sanctioned projects to be completed during the 12th Plan as no new sanctions are now being given. On normative basis, taking an average of 75 households per village, in all around 0.75 million households were provided with subsistence level lighting systems.[4] The MNRE is now considering revamping the programme to increase its coverage and effectiveness.[5] However, so far no scheme or policy has been declared. In mid-2012, the MNRE proposed a new scheme for village lighting scheme intended for electrified villages. We will discuss this in the next chapter.

The most surprising part of these figures is that Bihar is absent. Although the state was requested time and again, no proposals were sent to the MNRE. This was sad because the state had genuine problems of rural electrification. This was in spite of 90 per cent of the money being provided by the Central Government. This is difficult to explain and no one would be objectively able to find any rationale for this, except to conclude that the state has done well in village electrification or else there was no initiative taken in this area by the state administration. Jharkhand also faltered because the need was great, but perhaps the problem there was the inability to implement as time went on, although the start was good. In Tripura many projects were sanctioned and these were completed too. Assam was a very difficult experience as there were many and almost continuous implementation problems. To an extent, this was also true of Odisha. Chhattisgarh was the best and also had a different model including substantial financial contribution from the state and efforts to ensure sustainability which is discussed later in this chapter. In general, it was a difficult exercise and sustainability issues would only be natural.

In addition, there were two important nationally initiated special regional programmes—Arunachal Border Villages Lighting Programme approved in 2008 and the Ladakh Renewable Energy Initiative approved in 2009.[6] Under the Arunachal programme, 523 villages and 5,852 households were provided solar home lights. This process was completed between 2009 and 2010 even though it was arduous because of extremely difficult logistical conditions. It is hoped that these systems are operational even today and that the state government has taken steps to provide for replacement of batteries. There is yet perhaps no evaluation which needs to be done. In addition, a total of 132 mini hydel and small hydro projects covering 439 villages were taken up. In all, 95 projects have been completed covering 288 villages.

In Ladakh area of J&K, traditionally diesel mini-grids provided electricity for a few hours a day in many villages in remote areas which. Under the special Ladakh project, 62 40 kW/h solar PV power plants were proposed to be set up in both Leh and Kargil districts. These were all completed in 2012. Twenty-one small micro/small hydro projects totalling 11 MW are also under implementation in very difficult conditions in remote areas of Ladakh.

Incidentally both these are success stories. Credit must go to the state and local authorities. It shows that with determination and dedication very difficult work can be done. We need to replicate this spirit to cover energy access across India's plains!

Brief Assessment of RVE

In general, the RVE programme was quite successful in providing better lighting and replacing kerosene as the lighting fuel wherever it was implemented. This programme regularly went through the external evaluation process and the reports were generally encouraging.[7] But there were and are questions about sustainability of this programme besides there being many implementation challenges and issues. In some cases, particularly in Assam, there were cases of wrong beneficiaries. Theft of modules has been an issue. In many cases, particularly in Dhubri district, there were complaints of transfer and sale and export to Bangladesh. This could have been done by individual beneficiaries but could have also been partly an organised local racket. The fundamental problems related to maintenance and lack of service support. This had several dimensions. The most problematic is the issue of the battery and its functioning. In many cases beneficiaries replace the charge controller and connect directly. Early discharge is a problem. This reflects both poor awareness generation at the start and lack of general technician guidance. But it also reflects the tendency to misuse. The inability to change batteries and the failure of CFLs and difficulty in replacing them has been a problem. The absence of people from the supplier companies compounded the problem. Therefore, some systems have had different periods of workability as well as use for different hours per day. The evaluation reported many mounting and structural defects. One of the major barriers was that there were very few companies to implement. The main private sector company was Tata BP Solar which did a lot of work in Assam but had bad experiences there and thereafter abandoned this work concentrating on bank or CSR-financed systems. Reliance started with a bang in Arunachal and Tripura, but thereafter exited the solar power sector altogether. So the field was dominated by the two PSUs, Central Electronics Ltd and Bharat Electronics Ltd. A lot of effort was made in asking these and other organisations to open offices/dealerships in districts and to train local technicians, but this did not happen to the degree it should. This will remain a challenge for the future.

A CSE report that mentions findings of a team, which visited certain districts, lists some of these difficulties. Nevertheless, in spite of the difficulties, the report mentions about Bhetuli village 26 km from Almora in Uttarakhand, where BPL villagers were reluctant to get the grid-connected power because of high electricity bills. These villagers were also willing to pay certain instalments every month which after three years could cover the cost of the battery replacement. In Bung Bung village of Pithoragarh also, it is stated that the villagers when given an option between the capacity of the solar home system (SHS) being increased or being connected to the grid, the preference was for the former (Centre for Science and Environment, 2012).

The RVE programme by design fulfilled a short-term need and was not sustainable—very high subsidies; government-led; tender-based selection; absence of after sales service; lack of ownership; technical problems and an uncertain future because of battery replacement issues. But it responded to people's aspirations and reflected their hope and confidence in solar. Any model that is developed must resolve these technical, financial and structural problems while building on these aspirations by providing sustainable solutions. Later, we will look at what needs to be done now.

Bank-financed Solar Lighting Programme

While the RVE was being implemented in non-RGGVY villages, in the years 2008–09, the first serious attempts were made by some Grameen banks in collaboration with some solar manufacturing companies to give loans to rural households in electrified villages of different districts of Uttar Pradesh and Karnataka where the people were fed up with the unreliable supply of power and desperately wanted lighting in the evening.[8] Therefore, they were prepared to take loans and try out the new phenomenon of solar lights. These loans were without subsidy. In the next few years, many more banks and areas went through this exercise. It became clear from field visits to such villages made in 2009 that there was a great need for providing this facility to much larger numbers and that a proper programme was required to enable this to happen.[9] This led first to a programme with the banks themselves where the MNRE provided assistance for awareness generation and training. It was later developed into the very ambitious target of providing solar lights to 20 million rural households by 2022 under the Jawaharlal Nehru National Solar Mission through bank financing supported by government subsidy. Some may say that such a target would be in direct contradiction to the promise made of providing electricity through RGGVY to all households. Others, like us, argue that it reflects the ground reality that will not happen, and alternate solutions need to be found.

A special scheme was designed to provide both capital subsidy of 30 per cent of the benchmarked cost and interest subsidy so that interest charged was 5 per cent for balance loan portion for solar lights and small capacity PV systems up to 2 kW to rural households through banks including rural banks. However, the transaction costs and management of a dual subsidy regime became very difficult. Later, the interest subsidy was replaced and the capital subsidy was raised to 40 per cent. This is a package of two solar lights and a small fan and a point for television set. NABARD is coordinating the scheme.[10] Since the launch of the Solar Mission around 2.5 lakh solar units have been financed (Table 5.2).

These numbers are not very encouraging though it may be too early to judge the achievements of this programme. There is need for an urgent independent evaluation. This should examine the programme both when the banks were doing it independently earlier and later under the Mission with subsidy support and recommend steps required to upscale. A team from the Centre for Science and Environment (CSE) had visited certain villages in Barabanki and Hardoi districts where the Aryavart Bank has been lending since 2007. The team found somewhat similar technical problems as under RVE, lack of efficient after sales service and lack of

Table 5.2
State/bank-wise data of solar lighting units (as on 31 January 2014)

		Solar Lighting and Small Capacity PV Systems in Units					
S. No.	Name of the Bank	2009–10	2010–11	2011–12	2012–13	2013–14 (up to 31 January 2014)	Cumulative
I	**West Bengal**						
	Bangla GVB	0	244	456	4,654	1,136	6,490
	United Bank of India	0	0	0	1,787	417	2,204
	Central Bank of India	0	0	0	271	627	898
	State Bank of India	0	0	0	485	927	1,412
	Others	0	0	0	974	1,181	2,155
	Subtotal	**0**	**244**	**456**	**8,171**	**4,288**	**13,159**
II	**Andhra Pradesh**						
	Andhra Pragati GB	0	44	400	5,677	3,591	9,712
	Andhra Pradesh GVB	0	0	0	3,176	2,466	5,642
	Others	0	0	0	1,088	4,114	5,202
	Subtotal	**0**	**44**	**400**	**9,941**	**10,171**	**20,556**
III	**Rajasthan**						
	Baroda Rajasthan GB	0	0	122	92	10	224

(Table 5.2 Continued)

(Table 5.2 Continued)

			Solar Lighting and Small Capacity PV Systems in Units				
S. No.	Name of the Bank	2009–10	2010–11	2011–12	2012–13	2013–14 (up to 31 January 2014)	Cumulative
	Rajasthan GB	0	0	168	96	0	264
	Others	710	53	0	16	34	813
	Subtotal	**710**	**53**	**290**	**204**	**44**	**1,301**
IV	**Kerala**						
	North Malabar GB	0	0	0	18	11	29
	South Malabar GB	0	0	0	22	27	49
	State Bank of Travancore	0	0	0	61	79	140
	Others	0	0	0	26	757	783
	Sub total	**0**	**0**	**0**	**127**	**874**	**1,001**
V	**Uttar Pradesh**						
	Prathama Bank	8,758	14,669	9,763	8,459	4,463	46,112
	Aryavart Kshetriya GB	8,554	11,435	13,631	17,028	6,126	56,774
	Kashi Gomti Samyut GB	2,219	2,108	1,029	1,867	3,202	10,425
	Baroda UP GB	1,524	1,533	2,265	3,407	2,289	11,018
	Allahabad UP GB	0	7,343	5,511	1,158	1,260	15,272

	Others	10,922	2,515	2,148	1,071	1,604	18,260
	Subtotal	**31,977**	**39,603**	**34,347**	**32,990**	**18,944**	**157,861**
VI	**Tamil Nadu**						
	Pandiyan GB	0	0	39	33	0	72
	Syndicate Bank	0	0	0	32	0	32
	Others	0	0	0	16	57	73
	Subtotal	**0**	**0**	**39**	**81**	**57**	**177**
VII	**Karnataka**						
	Karnataka Vikas GB	8,007	8,006	6,168	0	2,863	25,044
	Kaveri GB	0	0	0	0	926	926
	Pragati GB	0	0	0	0	774	774
	Vijaya Bank	0	0	0	12	101	113
	Syndicate Bank	0	0	0	14	53	67
	Others	0	0	53	15	87	155
	Subtotal	**8,007**	**8,006**	**6,221**	**41**	**4,804**	**27,079**
VIII	**Bihar**						
	Uttar Bihar GB	1,027	1,373	197	149	0	2,746
	Others	0	0	0	43	641	684
	Subtotal	**1,027**	**1,373**	**197**	**192**	**641**	**3,430**

(Table 5.2 Continued)

(Table 5.2 Continued)

S. No.	Name of the Bank	Solar Lighting and Small Capacity PV Systems in Units					
		2009–10	2010–11	2011–12	2012–13	2013–14 (up to 31 January 2014)	Cumulative
IX	**Madhya Pradesh**						
	Narmada-Malwa GB	0	0	240	62	0	302
	Sharada GB	0	151	170	68	0	389
	Others	0	0	243	156	362	761
	Subtotal	**0**	**151**	**653**	**286**	**362**	**1,452**
X	**Jharkhand**						
	Vanachal GB	0	0	0	239	598	837
	Jharkhand GB	3,191	4,011	436	0	0	7,638
	Others	0	0	0	57	142	199
	Subtotal	**3,191**	**4,011**	**436**	**296**	**740**	**8,674**
	Other states	6,372	3,843	927	90	2,955	14,187
	Grand total	51,284	57,328	43,966	52,419	43,880	248,877

Source: Ministry of New and Renewable Energy.

ownership amongst beneficiaries. They mention that non-payment of instalments has become a major problem, although this did not generally appear to be the case earlier. One would have expected better service by the private supplier but it seems that there are systemic problems in servicing rural environments. Nevertheless, the team also noticed increasing demand at some places. The problem of battery replacement is yet to be addressed (Centre for Science and Environment, 2012).

It is evident from the above that the problems of electricity supply in electrified villages are real and that other solutions are possible. It also means that there may be more than 300 million people without real energy access. These solar lights were first provided without any financial support, and later at a subsidy of 40 per cent, much reduced from the 90 per cent subsidy under the RVE, and in already so-called electrified areas. This speaks volumes of what can be done with lesser resources.

Other Solar Home Lighting Models

World Bank experience: Let us briefly consider the experience and approach of the World Bank in this area.[11] Following the Rio Summit on Environment and Development in 1992, the World Bank started with a small solar loan to India in 1994. The implementing agency was Indian Renewable Energy Development Agency (IREDA), the funding arm of the MNRE. They did not perceive rural markets to be attractive for financing renewables like solar and they did not lend. A GEF grant had also been approved which would have reduced the rate of interest for the consumer to 2.5 per cent. The project did not take off.

Subsequently, the World Bank sanctioned loans to Indonesia and Sri Lanka for 200,000 and 30,000 systems, respectively, in 1997. The projects incorporated a grant per system installed (US$100) to the developer and a line of credit was made available which banks could on lend to the consumer. Preparations started well in the Indonesian project but the ensuing Asian economic crisis derailed its implementation. In Sri Lanka, there was success and installations by 2007 reached 100,000, with the help of a follow on project. The World Bank also sought to upscale the small Chinese market for small systems and approved a project for 350,000 systems. US$1.5 per watt was offered for any system above 10 W. The project was very successful and led to the installation of 500,000 systems by 2008 with 28 different firms participating. The average size of the system was small but gradually increased to 45 W.

In 2002, the World Bank also sanctioned a project for Bangladesh where Grameen Shakti, an NGO, had already emerged as a strong supplier of solar systems. The NGO had set up its own distribution infrastructure much like a business. The project enabled Grameen Shakti and BRAC to substantially increase the numbers. The project performed beyond expectations and, with assistance from other aid agencies, 200,000 systems were installed up to 2008. A target of 1 million was kept for 2012. The quantities have since increased substantially.

Essentially in these bank projects, entrepreneurs and new entrants to the market were explicitly supported with working capital loans and with grants and consumers were able to access bank finance. Wherever market infrastructure was created, the projects succeeded well. There was always an independent authority to facilitate and monitor the project.

Orb energy: This approach has led to a similar experiment in India, best reflected in the activities of Orb Energy out of Bangalore, set up by the author of *Selling Solar* (Miller, 2009). They are now a channel partner and get the subsidy from the MNRE directly. Thus, they are able to offer their products to the customers net of subsidy. We will discuss the issue of subsidy in Chapter 6. They have established retail stores in mofussil areas and are able to provide proper maintenance services through on call mobile technicians. This gives great confidence to the consumers and they are willing to pay higher prices. The question was how to make the additional costs of this nature of service viable. They were looking for small grants to set up a chain of such retail outlets in different states. They have now received a guarantee from US AID, which would help them to expand to 500 outlets. It would be appropriate for government policy to provide such support. The advantage of this model is that the systems will last longer and serve better and battery replacement will become easier later as it also would be facilitated. It is possible, in fact likely, that they would be able to serve the developing small urban areas too, where the electricity supply conditions may be as bad although they may not be able to reach the bottom of the pyramid because of higher costs. However, if there is a reasonable grant per system, the flexibility may allow them to balance supply of systems at lower costs to smaller clients with the larger clients paying more. There are other companies too that are trying to follow this model. Hopefully, they will be able to establish a proper decentralised eco-system, which unfortunately the solar companies/suppliers under the RVE or bank financed programmes have been unable to do.

Lighting a Billion Lives (LaBL): This initiative was launched by The Energy and Resources Institute (TERI) in the year 2007 with the aim of replacing kerosene-based lighting with solar lamps in un-electrified or poorly electrified villages. These micro solar enterprises were operated and managed by a local entrepreneur trained under the initiative. Although there were Solar DC Micro Grids, the primary innovation was the solar charging station model. A typical unit consists of 50 solar lanterns, 5 solar panels and 5 junction boxes. The solar lantern generates light for 4–6 hours daily, providing illumination of 200–250 lumens or the light equivalent to a 40 W incandescent bulb. These stations are set up in villages and lanterns are provided on a rental basis to households and enterprises every evening. To support sustainability of rural energy projects LaBL employed a network of local-level institutions to facilitate micro-implementation of project deliverables, carry out training and capacity building and ensure after-sales services. Under this initiative, 103,420 solar lanterns and 7,540 solar home lighting systems have been disseminated covering 115,834 households in 2,473 Villages in 22 states by March 2014.[12]

LaBL demonstrates how public–private people partnerships can support new initiatives in the area of rural energy access. It adopted a localised, bottom-up approach that has provided a valuable set of lessons and good practices. LaBL projects were funded largely by corporate or other donations and also received a portion as subsidy from the MNRE. To finance the campaign, TERI moved from the initial grant-based model to testing an entrepreneurial fee-for-service delivery approach, then gradually moving to a more flexible equity and investment-based model. However, it has been observed that these projects were essentially based on large subsidies and, therefore, in a sense it is not possible to replicate on a large scale, at least not immediately. The initial projects were also too spread out geographically and have a problem of sustainability. They would be better done in a cluster and would be extremely good for CSR projects. Schneider Electric Foundation has also implemented such a project in Sagar Island.

Programme for Left Wing Extremism districts: In view of the relatively small costs of a project and the need to meet basic lighting needs, on the recommendation of the MNRE, the Planning Commission included the Solar Lantern charging station programme in 60 (since increased to 78) Left Wing Extremism (LWE) districts in nine states.[13] It was suggested that they could preferably be done by women's self-help groups. The project provided 90 per cent of project cost as central financial assistance (CFA) and the balance 10 per cent was to be contribution from village entrepreneur or district administration (it should have been the entrepreneur only). In the first phase, 6,000 charging stations at the rate of 100 per district were approved. One 250 W peak (Wp) module was to be given to charge 50 LED lanterns and 10 mobile phones. The entrepreneur was to be selected by the district administration to manage the station by renting out the lanterns at nominal rate per day. The MNRE had tendered the project and shortlisted M/s Moser Baer, New Delhi and M/s MIC Electronics Hyderabad as vendors for the project in all nine states. Central assistance would be ₹1.35 lakh (90 per cent) and the remaining ₹15,000 by district administration/selected entrepreneur. Funds were to be released through state nodal agency of the respective state. So far, funds have been released to Andhra Pradesh, Chhattisgarh and Jharkhand. In Andhra Pradesh, out of two districts, installation has been completed in one district and in the other it is at an advanced stage of completion. In Jharkhand, the programme is under installation in half a dozen districts. In Chhattisgarh it is yet to take off. In remaining states of Uttar Pradesh, Bihar, West Bengal, Odisha, Madhya Pradesh and Maharashtra, the agencies are yet to identify the entrepreneurs. This is unfortunate and is perhaps an indication of the inclination of the state machinery or the Renewable Energy Agency concerned. We had been hopeful of better performance.

Mini-grid Model

Mini-grids are defined as one or more local generation units supplying electricity to domestic, commercial or institutional consumers over a local distribution grid. They usually operate in a standalone mode. There have been individual isolated projects from time to time in India and elsewhere. Diesel-operated mini-grids for villages or small rural markets have been existing for quite some time. The Village Energy Security Programme (VESP) of the MNRE was perhaps the first policy attempt to

Renewable Energy Options for Electricity Access 91

have community owned and operated mini-grid renewable energy-based systems. Thereafter, the Planning Commission thought of decentralised distributed generation as a possible model to supplement grid-based electricity supply to rural areas, which could be through diesel or any renewable energy source. The state of Chhattisgarh implemented RVE largely in this mode. There were some attempts by bilateral agencies. The MNRE also introduced mini-grid schemes which led to many initiatives by private entrepreneurs. Let us briefly look at these models and efforts.

Village Energy Security Programme: The MNRE developed the VESP in 2005 to go beyond electrification alone and address total energy needs (cooking, lighting and productive uses) of remote villages by using local biomass resources. It was considered a highly innovative and unique programme. It was also very ambitious. The World Bank provided technical assistance for 95 pilot projects in eight states of which only 51 per cent became operational. It was centrally sponsored and state administered while NGOs assisted programme implementation through Village Energy Committees (VECs). Electricity was to be provided by biomass gasifiers. A fuel plantation was to be developed although many locations being near forests fuel supply was not expected to be a problem. The programme did not succeed and had to be given up at the pilot stage but it provided many useful lessons and experiences.[14] It was a small scale but complex community driven programme in really remote villages. There was not much state level enthusiasm. Technology suppliers could not provide prompt and reliable after sales services in far off places and there were many small and big technical problems which resulted in disruptions. Local operators were inadequately trained and often did not get salaries in time. There was lack of organised supply of fuel wood. Individuals were to give their share. It did not happen. Biomass supply has to be monetised. Although the VECs showed great interest during planning and installation, there was lack of interest, as well as capacity, to manage the day to day affairs of the power plant (Figure 5.1). This suggests that plants be operated and maintained by dedicated and skilled local entrepreneurs who would ensure supply of biomass and collect dues from the consumers to whom they would deliver electricity services. (This is the Husk Power model.) It became clear that the revenues based solely on what domestic users paid (essentially kerosene replacement cost) was insufficient to cover the O&M costs. This suggests the need for

Figure 5.1
VEC model

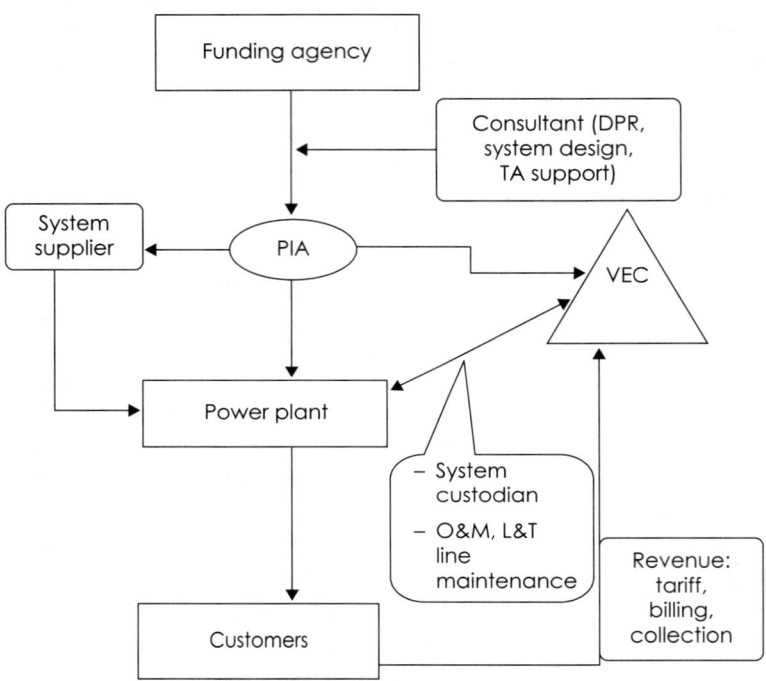

a well-designed viability gap support structure which would allow for a certain amount of profit. The mix of cooking and electricity provision also did not work leading to the conclusion that these should be separated and that a separate and dedicated cooking energy programme was needed. There were lessons regarding willingness to pay (WTP) and costs which will be discussed later. An interesting observation, tallying with that of RVE programme in Chhattisgarh, was that replacement of broken CFLs by incandescent bulbs was a big problem as it increased the load.

Decentralised Distributed Generation: In order to provide last mile connectivity as an adjunct to the RGGVY, the Planning Commission provided ₹450 crore in the 11th Plan period (2007–12) to the Ministry of Power for implementing DDG projects. The MNRE was of the view that mini-grids should be based on renewable energy only and this scheme should be implemented by it. This was not agreed to. Hardly any projects were implemented during this period. In spite of this, in the 12th Five Year Plan (2012–17), the allocation was doubled to ₹900 crore and again given to the Ministry of Power with Rural Electrification Corporation as the nodal implementing agency at the Centre. Clearly MNRE should have got these funds since it was implementing renewable energy mini-grid projects, and with a lesser amount of subsidy.

The state governments are to decide the Implementing Agency for this scheme. It could be the state utility or a central PSU or, ironically enough, even the State Renewable Energy Agencies. Projects are given 90 per cent subsidy of the project cost, of which 60 per cent is to be paid on commissioning and 8 per cent annually over the next five years. The project would be owned by the state government. After five years, the Implementing Agency could hand over the project to the same operator or another as approved by the state government for running the programme on a negotiated or limited or tender basis. So far, 719 projects have been sanctioned (Table 5.3).

Of these only 78 projects have been reportedly commissioned as of April 2014, and all in Andhra Pradesh, with very small solar PV mini-grid projects. On an average 96 W is being supplied for 6 hours per day to each household.

It may be seen that the on-going programme is a variant of the centralised grid with the main intention being that the utility should implement such projects. With such a liberal subsidy which is much more than that given by the MNRE, it is a surprise that very few projects have been sanctioned and even less implemented. Probably the transaction costs are too high. We will discuss this later, but the design of the scheme made failure inevitable. The persistence with this scheme and allocating more funds is indicative of the mind set to go the utility centralised way.

Chhattisgarh solar mini-grid model: Solar home lighting systems have been the primary means for providing lighting under the RVE programme. However, Chhattisgarh used these only for villages under 25 households and generally used the solar power plant and village mini-grid model. It also committed substantial state resources for meeting both installation and long-term maintenance costs. A strong service and maintenance network largely ensured the continued functionality of the systems.

Table 5.3
DDG projects under RGGVY (as on February 2014)

State	Name of Implementing Agency	Type of Project	No. of Projects	Name of District	Total Capacity of Projects (kW)	Villages/ Hamlets Covered (nos.)	Sanctioned Amount (₹ in Lakh)	Subsidy Amount (₹. in Lakh)	Subsidy Per Watt (₹ Per W)
Andhra Pradesh	EPDCL	SPV	251	Vishakhapatnam and Srikakulam	1,206	251	6,109.27	5,498.34	455.9
	NPDCL	SPV	39	Adilabad and Khammam	202	78	925.87	833.28	412.5
	CPDCL	SPV	9	Mahbubnagar and Kurnool	70	9	315.77	284.19	406.0
	SPDCL	SPV	11	Prakasam	143.5	11	469.44	422.50	294.4
Bihar	BSHPC	41 Hybrid Biomass gasifier + SPV & 7 SPV	48	Gopalganj and Kaimur	1,558	175	3,784.64	3,406.18	218.6
Chhattisgarh	CREDA	SPV	43	Surajpur, Balrampur, Korba and Sarguja	803	131	3,483.67	3,135.30	390.4
Jharkhand	DVC	SPV	43	Gumla, Hazaribagh and Koderma	518	89	2,090.91	1,881.82	363.3

(Table 5.3 Continued)

State	Entity	Type		Districts				
Karnataka	HESCOM, CESCOM, GESCOM and MESCOM	SPV	83	Uttar Kannada, Kodagu, Gulbarga, Madikeri, Mysore, Chikmagalur, Shimoga, Udupi, Dakshin Kannada and Chamarajanagar	877	193	2,977.86	2,680.07	305.6
	CESCOM	Wind & Solar PV (Hybrid)	10	Mysore and Chamarajanagar	138.2	11	612	550.80	398.6
Kerala	KSEB	SPV	15	Palakkad	131.5	18	531.83	478.65	364.0
Madhya Pradesh	MPUVNL	SPV	51	Sidhi, Umaria, Shahdol, Ashoknagar and Balaghat	882	162	4,265.53	3,838.98	435.3
Meghalaya	MECL	SPV	3	Jaintia Hills	154	3	389.40	350.46	227.6
Odisha	OREDA	SPV	7	Gajapati	65.5	7	197	177.30	270.7
Rajasthan	AVVNL	SPV	41	Udaipur and Pratapgarh	540.5	41	1,840.11	1,656.10	306.4
	JVVNL	SPV	1	Baran	14.5	1	51	45.90	316.6
Uttar Pradesh	DVVNL.	SPV	7	Hamirpur, Lalitpur	79	7	323.72	291.35	368.8
	MVVNL	SPV	14	Bahraich, Lakhimpur Kheri	973	55	3,733.93	3,360.53	345.4
	PVVNL	SPV	41	Sonebhadra	588	41	2,351.97	2,116.77	360.0
Uttarakhand	UREDA	Small Hydro	2	Tehri Garhwal and Rudraprayag	300	10	484.70	436.23	145.4
			719	**Total so far**	9,244	1,293	34,938.61	31,444.75	340.2

Since it is a unique experiment and has been in operation for several years it is worth a more detailed review.

The solar plant size was calculated with a provision of two lighting points for 6 hours per day per household and some street lights. The villagers selected the site and contributed paid labour for installation. The Plant was installed by one company and the distribution done by another. Ninety per cent payment was made after installation and balance 10 per cent after a period of five years. O&M was outsourced to another company based on annual contracts with a cluster service model of 10–15 villages. After installation, the system was handed over to the VEC, although the main task remained with the O&M Company. The work started with extensive training to the technicians and the person who would be the operator. There were connection charges of ₹100 for the BPL household, ₹200 for non BPL and ₹500 for others. Maintenance charges were ₹30 per household per month of which ₹25 are paid by the state government and only ₹5 by the beneficiary. Although this ensured that the systems were not dependent upon customer payments for their maintenance, perhaps this charge for the beneficiary was kept at too low a level as it is not sustainable in the long run. More importantly, it gives the message of power supply largely being a dole. Funds for CREDA came from taxes levied on electricity bills paid by all customers. This is a good idea, but perhaps it should come as a surcharge on the larger customers based on an incremental tariff structure. The O&M Company gets ₹25 per working connection.

There was a study for a PhD thesis regarding the functioning of these systems 2–3 years after installation (Milinger and Marlind, 2010). Some interesting observations have been made therein. It was noted that there were several initial challenges of which low awareness of people; dispersed nature of population; difficulties of road access particularly during monsoon; and unorganised service were among the most prominent. All these increased costs. It is important to note that these are not captured when subsidy levels are calculated by the MNRE. However, it is hoped that these might ease with time as more villages are electrified in this manner and rural customers grow. It was found that SPV power plants were 37–80 per cent more expensive than SHS for the same hours of light, although their service was better, because of the operator; there was less damage to bulbs (40 per cent for SHS and 20

Renewable Energy Options for Electricity Access 97

per cent for grid); batteries were serviced better. They also gave light for 6 hours as against 3 for the SHS. Willingness to pay (WTP) was higher for SHS at ₹19 per month as against ₹10 for the other. Actually, with two litres of kerosene being saved per month and its cost being ₹15 per litre, the WTP should be higher. SHS led to a larger reliance on kerosene during monsoon. SHS used on an average 3.1 litres of kerosene during the month which usage declined by 2.1 litres. The same figures for grid were 3.6 and 2.6, respectively. Although problem of overload is quite significant, it is more in SHS, because of use of appliances, mostly TV. It was strongly suggested that the initial capacity be much more than what was set up. This happens perhaps because of initial over calculation of efficiencies or underestimating losses. This factor also needs to be kept in mind in the design of the subsidy. It is suggested that a village with about 125 households which got a plant of 6 kV capacity should have actually been double that. It was also observed that 10 per cent of the installed inverters were damaged every month. This is a very serious issue as it increases the costs of maintenance and disrupts supply affecting both recoveries and confidence/credibility. The thesis quotes a study from Mexico on 555 lead acid batteries which showed that undersized storage capacity was one of the main features of contributing to battery capacity degradation. Overload is also caused by changing from CFL to incandescent bulbs. It is suggested that there should be both a subvention for better lighting systems, like LED or CFL, and a market ensuring easy availability for their replacement. All this, and use of TV sometimes, also leads to diminishing battery capacity.

It was found that the role of the VEC at many places was not very constructive. Efforts at improving this area have to continue. It was also found in general that involvement of women was not very much. Therefore, there would be a need to involve the women's self-help groups. However, women clearly were better off. Since this is a valuable benefit, it would be useful to quote the observation in the thesis:

> The women clearly obtain a benefit as long as the power is provided and works reliably, both by their own statement obtained through the conducted surveys, but also judging on the changes in habits in cooking and the willingness to pay. They do not have to plan the daylight hours according to when they have to cook, and they do not have to rush the cleaning of the utensils and the households after eating before sunset.

This provides both a reduction in stress, but also an increased hygiene and prolonging of the time that they can work in the field or with productive activities. It increases the choice they have and their adaptability to unforeseen events, which otherwise might disturb their routine before sunset. However, this only applies as long as the light in the evenings is reliable to them. Once it passes a threshold of unpredictability, the habits must return to adapting to the time of sunset regardless of if there is light on that particular day.

The study has made some recommendations for CREDA and other agencies working with remote electrification. These would be generally useful.

- Install sufficient power capacity as it would reduce conflicts in the villages as well as maintenance costs;
- Consider also homogeneity and cohesion in villages when choosing what type of system to install;
- Higher demands on installation firms and more focus on maintenance of wiring;
- Install more robust inverters with timers;
- Focus on implementing a Remote Monitoring System as it would increase transparency and focus attention on the villages that need it;
- Older, more respected people as operators;
- Better possibilities for VEC to change operator;
- Clearer responsibilities for technicians, operators and VEC members;
- Include provision of CFLs/LEDs into CREDAs sphere of responsibilities; and
- Cooperate with other governmental schemes and agencies involved in the remote locations.

Projects of Bilateral Agencies

GIZ project: In 2007, in collaboration with partner organisations in India and Germany, GIZ implemented a project to facilitate access to biomass-based electrification and energy services in rural India. The project focussed on reviewing and assessing the current practice of sourcing and the supply of locally available biomass resources, as well as the use of by-products. Initially 20 pilot projects were planned. However, only 10 plants were set up, and of these, only three projects are stated to be functioning. These plants adopted the biofuel and biogas route. As such the project could not achieve its stated objectives.[15] It would have been better if they had been planned together with the MNRE. In any case they serve to reinforce the earlier observations on the VESP Project.

Indo-Norwegian project: The Government of Norway, the MNRE and a Norwegian Company, Scatec Power, collaborated and co-funded a solar-based mini-grid rural electrification project in 2009–10. This aimed at developing sustainable and scalable business models for accelerating large-scale roll-out of community solar power plants in India. Pilot projects were set up in 28 villages in four different states in India, namely Uttar Pradesh, Madhya Pradesh, Jharkhand and J&K. Scatec Solar worked with local NGOs, who acted as door openers into the project villages. The NGOs worked to mobilise the villagers and analysed the needs of the community. These models were based on the idea of providing solar power 'beyond the light bulb' for some commercial activity also. The NGOs worked with the villagers to promote the development of income generating activities which may take advantage of the arrival of electricity to the village. Under this Indo-Norwegian cooperation project, 28 villages have been electrified with MNRE contribution of ₹5.4 crore and Norwegian commitment was ₹12.3 crore. Solar PV capacity of 280.60 kWp solar PV was installed. Apparently, however, the electrification cost was exorbitantly high. This has not been shown as a sustainable business model. Nevertheless, there are useful lessons to be learnt. Scatec Solar has summed up these as under:[16]

- Selecting villages above a certain size is essential as there are prohibitive fixed costs for the smallest village installations;
- Battery banks are costly, combining PV with a second power source in a hybrid set-up is preferable;
- Using existing grid infrastructure would help keep cost down (e.g., mini-grids for DG sets);
- Dimensioning of the PV system is crucial—but very complex in a greenfield setting;
- Investment support is needed, preferably also subsidising the power sold to the end-user, not merely the upfront capital investment;
- Proving guarantees for off-take is required, e.g., a first loss mechanism;

- Systems should first be introduced where the alternative cost of energy is high;
- Mobilisation and motivation of NGOs and villagers is key;
- Tariffs should be dynamically set (e.g., higher tariffs at evening);
- Price discrimination should be deployed where commercial off-takers are able to pay a higher price for power than residential consumers;
- Commercial loads are important for these types of systems. Purely domestic off-take could often be better served by other systems and business models (micro financed, self-owned home systems);
- NGOs would need to take full ownership and be actively promoting optimal use of the systems; and
- Geographically, all installations should be clustered in one homogeneous location.

MNRE mini-grid scheme: In 2009–10, the MNRE had a scheme for mini-grid which provided a support of around ₹6 lakh for a 10 kWe capacity through biomass gasifier and around ₹3 lakh per kW for solar mini-grid projects and an additional support of ₹3,150 per household towards the cost of distribution lines, service connections, fittings, etc.[17] There were many innovative efforts using biomass like Husk Power Systems and Saran Energy in Bihar. Many years earlier, solar grids had been set up in Sagar Island of West Bengal. These are discussed in some detail below.

In the meanwhile, it is heartening to see many entrepreneurs entering this area with small pilot mini-grid projects. Gram power is an energy technology company started out of University of California, Berkeley. It primarily focuses on rural electrification with use of cutting-edge smart Grid technology. In little over a year of commercial operations, the company targeted 21 off-grid villages to be electrified in Rajasthan and Uttar Pradesh with Solar Powered Smart Micro grids. One of the key features of Gram Power's smart meter is that it works on the pre-paid model analogous to pre-paid phones. This ensures that sold power is completely paid for and also saves the interest for the two months of billing cycle which is the current process for Discom. Moreover, each unit of power consumed is accounted for in the multi-level architecture. As the payments are wireless and prepaid, collection costs are reduced significantly. The system consists of a 240 V AC connection for two CFL lights at ₹11,500/family. The smart grid pre-paid metres allow power supply only on prepaid basis and can be recharged for ₹20 only. It claims to meet the three main risks for mini-grids—theft, payment and affordable reliable supply.

Another example is of Omnigrid Micropower Company Pvt Ltd (OMC). In 2010, three former Ericsson employees teamed and floated OMC Power that is beginning to change the rural landscape, both social and economic, in Hardoi district of Uttar Pradesh. Villages covered by it are reaping the benefits of reliable, renewable and cheap electricity. Typically, OMC Power sets up 10 kW solar power plants. The power generated is used to charge solar lanterns and batteries that are rented to homes through a channel of dealers. In a single charge, a battery weighing 2 kg lights up household equipment, including a fan, a bulb and a television. The battery, locally called the bijli box, is replaced by a charged one every evening by the company's employees. OMC Power charges ₹120 a month for the lantern and ₹250–350 a month for the battery. OMC Power operates 10 power plants with a total capacity of 100 kW. It lights up about 5,000 households in a 30 village cluster 70 km away from Hardoi town. It has been estimated that based on this model around 20,000 plants could be set up.

Mera Gao Power (MGP) builds and operates 240–300 W 24 volt DC micro grids in Uttar Pradesh. It focuses only on basic lighting and mobile point. It claims to have the lowest cost micro grid. It charges a monthly fee of ₹100. It has up-scaled to 13,000 customers in a large number of villages. It says it is meeting maintenance cost through revenues earned.

There are many other small initiatives by different individuals and organisations. The Solar company Sun Edison is doing a few villages under its CSR programme. It has a 14 kW model designed for about 100 W per household for nearly 8 hours daily. Minda NextGen Tech has started in a few villages in Uttar Pradesh and wants to do 100 in the next couple of years. They set up the system and have a local entrepreneur. In the CSR system, each household pays ₹25–50 and the money collected will be used for change of batteries. In the entrepreneur model, the households pay ₹75–100 and the battery would be changed by him. These are small projects covering only 40 households each with the system cost being ₹1.75 lakh.[18] The Company takes responsibility for the systems and battery working for 5 years. The Freemasons want to fund 50 villages across India. Gram Oorja is doing several villages in different areas. It would be useful for the MNRE to prepare an inventory of all the projects and put the main features of all projects on its website. There is a need to learn from these efforts and experiences and create the synergy necessary to upscale.

Sagardeep Island, West Bengal

Sagar Island is home to around 200,000 people. Prohibitively high costs have kept it disconnected from the grid. The West Bengal Renewable Energy Development Agency (WBREDA) became interested in Sagar Island in 1994, and has since used both solar PV and wind–diesel hybrid systems to electrify much of the island. The programme began with the installation of individual solar lighting systems, which provide electricity for more than 2,000 families. There are 10 SPV power plants carrying a total capacity of more than 300 kW and powering more than 1,600 households for six hours a day. Beginning in 2002, WBREDA initiated a wind-diesel hybrid programme. The system consists of two 180 kW diesel gensets connected to four 55 kW wind turbines. Whenever the turbines cannot generate enough electricity to meet demand the gensets make up the difference. This system now meets the demand of more than 700 households and major commercial parts of the island. WBREDA has also added biomass capability to its hybrid system further lessening the need for diesel and increasing the power. The project was designed to bring electric power through local mini-grids.[19]

Renewable energy interventions in Sagar Island are community managed. The Sagar Rural Energy Development Cooperative (SREDCOP), composed of important local officials, oversees all renewable electrification on the island. Each power plant then has a beneficiary committee, made up of a SREDCOP member and other village leaders as well as consumer representatives, that is responsible for consumer selection, setting and collecting tariffs, and some aspects of maintenance.

Fifty per cent of capital funding for the solar mini-grids was provided by the MNRE, another 20 per cent from the provincial government, both in grant format. The remaining 30 per cent came as a soft loan from Indian Renewable Energy Development Agency (IREDA) under the World Bank assisted SPV market development programme and revenue from consumers. Customer revenue covers operating costs. The plant charges its customers a tariff based on the number of connected power points rather than the actual amount of energy used. Consumers pay ₹85/month for 3 lighting points, and ₹135/month for 5 lighting points. There are also a few bulk customers who pay a fixed rate. The power is mainly used for lighting and the running of small appliances, machinery and radio/TV. The total revenue of the plant at approximately ₹1.5 lakh per year (corresponds to an average payment of roughly ₹100/month). Each household needs to pay only once a month and there is a near full payment rate. The high compliance is credited to the managers' close ties to the community, which provide them with detailed knowledge of their consumers.

The increase in electricity access in Sagar Island has brought attendant economic, social and health improvements to the island as well. Two hospitals run on solar energy, and 80 streetlights installed around the island enable residents to travel and congregate more safely at night. Lighting the streets has led to the creation of small businesses along the island's main thoroughfares whose profits grew based on increased hours of operation.

The Sagar Island project is not a business model. The existing revenues are just sufficient to meet current operational cost requirements. The high visibility and aggressive support from WBREDA enables these plants to continue to run. The expenditure on major repairs and battery replacement are met by WEBREDA as and when required. Thus, this model is highly government support driven and its replication is directly linked with the continuous attention and assistance from the supervising state agency. Community management systems, however, have not created the incentives necessary for profit-maximising behaviour amongst the plant's local managers, with negative effects on sustainability. Moreover, without a direct financial stake in the project, the local community lacks the incentives to create a self-sustaining system. A more profit-oriented business model that reflected real value tariffs and the cost of alternatives would require professional management that may have more direct incentive to ensure that all available demand was being met, by expanding generation if necessary.

Husk Power Systems for Electricity Access, Bihar

To the non-resident Indian (NRI) Gyanesh Pandey and his friends go the credit of pioneering one of the most exciting local electrification efforts in power-starved Bihar. He used the plentiful rice husk available locally in villages, much of it being transported to other states. Rice mills had been using dual fuel gasifiers with producer gas and 35–40 per cent diesel. This was not possible for rural village supply. He was able to use gasifier in single fuel mode. After a small beginning, this technology is now meeting the electricity requirements of around 250 villages in Bihar leading to cost savings for households buying electricity from this source replacing kerosene use as well as considerable opportunity for job creation.

Husk Power Systems (HPS) is based on a proven biomass gasification technology of standard fixed bed, down draft type, which is suitable for rice husk-based power generation of a capacity range below 200 kW. The capacity of the systems varies from 25 to 80 kW with the cost of a 32 kW system being ₹1.6 million. The MNRE subsidy works out to around 50 per cent of the total project cost and is directly transferred to local entrepreneurs in the case of BOM and BM business models and kept by HPS in case of BOOM model. The rated capacity of the typical HPS systems installed in Bihar is around 32 kVA. These generate nearly 25 kW of electricity of which 1–1.5 kW are lines losses; average theft amounts to around 1 kW; and approximately 18–19 kW is available for consumption or saleable purpose. While every household has to pay a fixed monthly charge of ₹45 per CFL of 15 W, shops pay a per month charge of ₹80/CFL. There are different payment categories for other productive applications. HPS has adopted a demand driven approach and quantifies the potential demand in watt-hours, minimum to be around 500–650 households would be there. Households are required to pay installation charge of ₹100. This money ensures compliance by the users, and also covers a substantial portion of grid distribution expenditure. HPS ensures that the power plant machinery meets the requirement of the power plant. It has tried to reduce costs by procuring gasifiers from local manufacturers and coupling it to modified gas engine enabling operation on 100 per cent producer gas.[20] Collection is about ₹40,000 per month and expenses about ₹25,000. Therefore, it is financially viable. There is enough profit for a small entrepreneur, but not enough to drive large investments, unless there is a large cluster. There are also risks.

HPS manages most of the decentralised power-generating units with the help of four personnel assigned to each plant. For the day-to-day management, every power plant has one operator and one husk loader, where in the operator carries out the routine maintenance.

Renewable Energy Options for Electricity Access 105

The operators are trained by HPS in Patna (Bihar) for two months and then sent for on-the-job training in one of the operational plants. In addition, two more people are associated with each plant. One handles husk buying and ensures a regular supply of raw material and is also involved in the revenue collection activities. The other is an electrician for a cluster of villages. In addition, HPS has cluster level manager who looked after the plants in the range of 20–25 km or about 5–7 plants. Besides trained manpower, HPS has also taken due care to ensure smooth supply of low-cost raw material. At the market end, the promoters have evolved strong relationship with the rice husk suppliers. This husk is transported by tractors simultaneously to about 7–8 plants in one cluster.

Husk Power made many innovations and their business model has many competitive advantages. These include

- managing the process from beginning to end—from power generation to maintenance and revenue collection;
- investing in research and development (R&D) and establishing strategic partnerships to design simple equipment to keep costs low and localise gasifier/engine production;
- improving distribution efficiencies including developing an instrument to prevent theft and increasing value for everyone;
- creating jobs for local people who are trained and employed to operate the power plants and receive a market rate income without having to immigrate to the cities, helping to create local buy-in;
- setting up a training school for technicians;
- at least a 35 per cent cost saving for households and businesses that receive power from Husk by replacing kerosene lanterns or diesel genset-based lighting; and
- using by-product as ash ball for local manure or cooking or for being converted to *agarbattis* by local women and sold to India Tobacco Company (ITC).

All these efforts made the system become more viable with the subsidy support of the MNRE. But Husk Power has faced difficulties, many of them local. A major problem has been delays in the release of the subsidy. It seems that there have even been unrealistic return expectations of the investors. Husk Power realised that the management of multiple units by a single entity is not a cost-effective proposition. In order to create ownership on the ground, they started 'franchising' the plants, where a local entrepreneur invests in the plant machinery and distribution infrastructure. This model has shown success but the real challenge lies in capital availability for such entrepreneurs. Nationalised banks have not been very forthcoming in financing such projects. Few branches of Central Bank of India have started lending entrepreneurs under the Credit Guarantee Trust for Medium and Small Enterprises (CGTMSEs) scheme but the progress has been rather slow. Banks have also been concerned about the unpredictability in release of subsidy from the MNRE. Delays in the release of subsidy leads to piling up of interest on the entrepreneur and stresses the business.

It may be observed that biomass gasifier grid requires management and technical supervision of a much higher order than a solar grid. Its feasibility is very site specific due availability of feed material and divergences in prevailing fuel prices. Therefore, upscaling of biomass gasifier-based electrification projects require a different kind of local entrepreneur.

Saran Renewable Energy

Saran Renewable Energy (SRE) was set up in 2006 by Mr V.K. Gupta as a small family-owned firm. SRE has built a biomass gasification plant at Garkha, 70 km from capital Patna to gasify biomass purchased from local farmers to generate electricity which is sold to local businessmen.[21] The gasifier installed at Garkha by SRE is designed to supply 128 kW electricity at 240 V, a high voltage for a gasifier with two 3 kV transmission lines, each 1.25 km provide link to the customers. The plant is run for 10 hours every day using 35 per cent of the total capacity. Three gasifiers have been built. The gasifier used by SRE is down-draught-open-top gasifier. The power generated by the SRE plants cater to about 1,000 households and many businesses and shops, two grain mills, a cold storage unit, a cinema hall, a saw mill, two medical clinics and several genset operators. Most of the genset operators are those who were earlier running diesel generators to sell electricity in the villagers. The gasifier has not affected their business as they now sell electricity bought from SRE and retail it to households and business establishments.

The sale price of electricity to the consumer is ₹7.5 per unit. The cost is calculated taking into account the pay back of the loan taken by SRE to set up the gasification plant. Although the cost by the State Electricity Board is pegged at ₹6 per unit, customers are willing to pay a little extra for reliable supply, stable voltage and higher frequency. The charges for power from diesel generator are ₹12–16 per unit. A meter attached to the supply of each customer records the consumption every day. If the payment is late by more than a week, the supply is disconnected. The gasification plant operates daily for about 10 hours from 1,000 to 2,100 hours (1-hour break) with a current peak demand of 90 per cent of the capacity with an average demand of 65 per cent. Twelve staff and five casual workers run the plant.

The cost of the entire system was ₹8,300,000 with the cost of the gasifier and generation plant around 90 per cent of the total cost and the remaining 10 per cent for the two 3 kV distribution lines. The plant was set up with equity participation of owners and ICICI Bank loan of 20 lakh. A government assistance of ₹18 lakh was also received.

Although the company recorded a loss in the first year of production, it was able to make a profit of ₹6 lakh in 2008. In the break-up of the expenses incurred, the largest portion goes towards paying off the investment (55 per cent), fuel cost (35 per cent) and operation and maintenance (15 per cent). SRE buys agricultural waste from nearly 100 farmers. This is called *daincha*, which grows locally on the marshy land. The price varies between ₹1.5 and 2 per kg depending on the moisture content. Framers thus gain an additional income of ₹7,500–10,000.

Business of many of the customers of SRE has improved. They include grain and oil mills, a saw mill, a welder, a battery charging station supplying lighting. All benefit from an increase in their business from a reliable supply of electricity. Farmers living close to the transmission line use electricity to operate about ten irrigation pump. A blood collection lab now can work unhindered by paying ₹200 per day as compared to 300 per day for supply from a diesel supply.

The economic life in that area has been transformed. There is a lot of potential especially for areas which have a cluster of small businesses located in areas where such biomass can be easily obtained. In this context, Garkha plant is a business model and its success is primarily linked with the affordability of customer and paying capacity of people to the competing option. However, there are not many such plants and the success is limited to a few clusters in Bihar where there is a case of high premium on electricity for productive applications. This model has substantial replication possibilities but cannot be considered a representative model for electricity access projects.

Lanterns

There is a huge market developing for solar lanterns of various sizes and strengths. There are now over 50 manufacturers. By December 2013, around 1 million solar lanterns had been distributed in the country (Table 5.3). It would be difficult to gauge the numbers of lanterns sold in the market but they could be substantial. They may also all not be of good quality and there are problems of maintenance. In the last few years, the MNRE only sanctioned some special projects for lanterns like the shepherds of Kashmir/border districts of Uttarakhand and in other difficult regions of the country. In view of the difficulty of ensuring their sustainability, it was felt that this mode was best left to market forces.[22] But some state governments have been distributing lanterns although they are no longer entitled for central subsidies. Chhattisgarh is currently implementing a big programme which includes small study lamps for children who have secured positions in school examinations. There is no evaluation done regarding the present status of the lanterns distributed in earlier years. Some must be working with the original batteries supplied. In some batteries may have been changed. Others may have simply been given up or the solar panels being used in some other way. Unless there is a real decentralised ecosystem for maintenance and supply of parts or batteries, etc., this is not a system in our view which should be encouraged by government supplies. On the other hand, since it is a low-cost product, the development of an ecosystem should become a priority (Table 5.4).

Table 5.4
State-wise cumulative solar lanterns (as on 31 December 2013)

S. No.	State/UT	Lanterns
1.	Andhra Pradesh	41,360
2.	Arunachal Pradesh	14,433
3.	Assam	1,211
4.	Bihar	50,117
5.	Chhattisgarh	3,311
6.	Goa	1,093

(Table 5.4 Continued)

(Table 5.4 Continued)

S. No.	State/UT	Lanterns
7.	Gujarat	31,603
8.	Haryana	93,853
9.	Himachal Pradesh	23,909
10.	J&K	44,059
11.	Jharkhand	23,374
12.	Karnataka	7,334
13.	Kerala	54,367
14.	Madhya Pradesh	9,444
15.	Maharashtra	68,683
16.	Manipur	4,787
17.	Meghalaya	24,875
18.	Mizoram	9,589
19.	Nagaland	6,766
20.	Odisha	9,882
21.	Punjab	17,495
22.	Rajasthan	4,716
23.	Sikkim	23,300
24.	Tamil Nadu	16,818
25.	Tripura	64,282
26.	Uttar Pradesh	62,015
27.	Uttarakhand	64,023
28.	West Bengal	17,662
29.	Andaman and Nicobar	6,296
30.	Chandigarh	1,675
31.	Delhi	4,807
32.	Lakshadweep	5,289
33.	Puducherry	1,637
34.	Others	125,797
	Total	939,862

Source: Ministry of New and Renewable Energy.

Franchisee Model

We discussed a model in the previous chapter based on utility supply of power. There has been some discussion on a suitable franchisee model based on decentralised generation. The World Bank had also proposed this. It did a study analysing its viability (World Bank, 2010). An average rural household spends almost ₹11/kWh to meet its lighting needs. The average retail tariff is between ₹3–4/kWh for domestic rural consumers. The cost of supply is taken at about ₹7. If the utility appoints a distributed generation and supply operator, the gap is likely to be about ₹4/kWh. Surveys by TERI in 2010 in Ding (Haryana) and Radhanagri (Maharashtra) showed that more than 80 per cent of the households surveyed were willing to pay additional amounts of ₹1.60–3.20/kWh over the existing tariff of ₹3.64/kWh and ₹1.40–3/kWh over the existing tariff of ₹3.77/kWh, respectively, for improved power supply. But while the willingness to pay for initial lighting needs is very high, it declines progressively with each additional kWh. The World Bank has calculated the average cost of supply for Ding and Radhanagri at ₹4.49/kWh and ₹4.94/kWh, respectively, in case of utility supply. With collection efficiencies of about 90–95 per cent the tariffs to be levied would be ₹4.72 and ₹5.20, respectively. A local biomass power generator would be willing to adopt the additional role of distribution franchisee if it yielded a tariff of ₹7.11/kWh or average revenues of ₹6.76/kWh at a collection efficiency of 95 per cent. The losses for the utility are seen as ₹1.27 and ₹1.37, respectively, which the World Bank suggests the utilities pay. The balance ₹2.39 and ₹2.60 would need to come as viability gap funding from some other source.

The question is: why has there been no movement in this direction? It is essentially because to create this model, there are a host of other conditions which are considered difficult to fulfil. Financial modelling for undertaking investment decisions in a situation where tariffs and other conditions may change from time to time is really not possible. Besides, there is dependence upon the utilities, even for payment. This is a big risk, especially for small units. There are also hidden costs and transaction difficulties. This model also does not address the issue of the losses to the utilities because they continue to pay the difference of the cost of supply. At the same time, the analysis shows that the consumers will be paying a bit more than what they currently pay for uncertain electricity, especially for lighting. Therefore, we are actually establishing the case for an independent off-grid supplier of power through renewable energy with proper viability gap funding. No wonder even the World Bank is now proposing such a scheme for a cluster of villages in Bihar and Uttar Pradesh. The franchisee model is dependent upon a properly functioning grid, a generally reliable infrastructure of supply, a supportive utility, a rationalised tariff, etc. None of these apply, in fact to the contrary. This is also the general conclusion of different analyses on this issue.

Localised Grid-based Supplies

There is an alternative model to strengthen supply of energy to rural areas. This is simply to feed renewable energy through 500 kW to 2 MW plants into the 11 kV grid. This is possible through either solar or biomass. These have the following advantages than feeding large amounts of renewable power into the grid at faraway places like solar and wind in the desert areas of Gujarat and Rajasthan, or the proposed solar mega plants in Ladakh or the salt plains of Rajasthan.

1. Power generated almost at consumption point.
2. Likelihood of power being used for adjoining rural areas.
3. Saving of transmission loss, possibly 7 per cent.
4. Quality of power improves.
5. Power is available in daytime, and in case of biomass, for the evening peak as well.
6. Short gestation periods.
7. Lesser requirement of funds for setting up plants.
8. No need for building huge transmission/substation infrastructure.
9. Land problems may be considerably eased.
10. Smaller local entrepreneurship may develop because of smaller plants.
11. In case of biomass, supply will help raise incomes of farmers in nearby areas through supply of waste biomass. Fuel supply in such quantities may not be a constraint.

12. There are possibilities of growing dedicated plantations of fast growing bamboo or tree species in nearby waste lands or degraded forest lands, which would add to rural development in that area.

Some Examples and Models

Malavalli Biomass Power Plant

Malavalli Biomass Power Plant in Karnataka has operated since 2001, with a power evacuation facility to the Karnataka Power Transmission Corporation Limited grid at 11 kV, which continued to provide electricity when the main grid was not there. To help support residents, Grameena Udyog Samiti (GUS) was established as a platform for local farmers and an NGO was set up, Grameena Abhivrudhi Mandali (GAM), which manages as a franchisee power distribution, billing and revenue collection on behalf of the distribution company. Procurement of biomass fuel from local farmers and biomass suppliers has generated additional income and improved economic condition of the community. This has also resulted in local employment generation. Plant has generated employment opportunities directly/indirectly to more than 400 people. As a part of social responsibility, plant has been contributing to social infrastructure by way of employing local people for the plant operations and also paying significant amount as tax for the local panchayat, etc., surplus crop residue is used for power generation, which was otherwise burnt in the fields. Project activity resulted in generation of direct and indirect employment due to biomass collection, transporting and unloading, etc. (World Bank, 2010; Sutter, 2013). There are good opportunities for such projects, but the tariff paid by the utility would be the main constraint. This project was also a successful example of utilising carbon credit. Although a difficult model, it is of great value and worth replication.

One Megawatt Wood Biomass-based Gasifier Plant in District Tumkur, Karnataka

UNDP/GEF set up a project for 1 MW of a wood biomass-based gasifier plant in District Tumkur, Karnataka, a few years back also for feeding into the 11 kV grid substation. This was an integrated project and included growing fuel wood plantation on 3,000 ha of land, some biogas plants and bore wells for irrigation. The fuel wood yield was 5,000 T against expected 12,000 T. This is remediable. The plantations were fragmented in nature, which led to increased transportation costs. The capital cost of the plant was ₹7 crore. The fuel cost was 57 per cent of the running of the plant, which could have been less. The generation cost came between ₹4.50 and ₹8.28. The grid tariff was only ₹2.85. The economics of the project could improve greatly, but eventually the tariff paid will be critical. An entrepreneur is setting up another plant based on such plantations near Madurai in Tamil Nadu, but he struggled to get funding, and other clearances. Such plants based on dedicated fuel plantations need to be encouraged and paid proper tariff or get a viability gap funding. We had suggested in Chapter 3 that a capacity of 20,000 MW providing base power could be developed over a period of time.

1.25 MW Biomass Gasification Power Project Located in Coimbatore

Arashi Hi-Tech Bio-Power Private Limited (AHBPPL) established a 1.25 MW biomass gasification power project located in Coimbatore district, Tamil Nadu. AHBPPL was the first grid connected biomass gasification power project in India supplying power to Tamil Nadu State Electricity Board (TNEB) grid. The fuel for this power plant was locally available biomass residue like coconut shell. The income of local people increased by selling biomass residues to the power plant. However, the plant has not been operating anywhere near its full capacity on account of drastic increase in the price of coconut shell which was the main raw material being used.[23]

1.2 MW Grid-connected, Biomass Gasification Power Plant in Sankheda Taluka, Gujarat

Ankur Scientific has set up a 1.2 MW grid-connected, biomass power plant based on its own gasification technology in Sankheda Taluka of Vadodara district. This project is the first of its kind in Gujarat and also the first project to be set up under the status of 'Model Investment Project' implemented by the MNRE and United Nations Development Programme

(UNDP) with partial financial assistance from both. This project has been registered for availing Renewable Energy Certificates and the electricity generated from the said power plant is supplied to M/s Aditya Birla Insulators under open access provision. This appears to be a viable model because of better tariff. The project needs to be studied in detail. Fuel supply is reportedly manageable and local incomes have increased.

Avani Initiative of Pine Needle-based Power Plant

A unique plant of 120 kW is being set up in Pithoragarh district of Uttarakhand by Avani, an NGO working in that area based on pine needles as fuel. They got funding through Acumen Fund, facilitated by MNRE, but with great difficulty. A new company had to be formed. It took 2–3 years for it to go through. It is expected that the plant will be commissioned by mid-2014.[24] Avani was already running a 10 kW plant. This has many advantages. It is environmentally friendly since it removes pine needles and therefore the real threat and danger of forest fires, which have actually happened. It gives some incomes to those who collect pine needles. The residues can help provide cooking fuel to 100 families. And the rural grid is strengthened. The cost of the plant is about ₹70 lakh. There are opportunities for replication in all the hill states and several hundred plants can be set up. However, the key will be tariff. The hill states having hydro power are reluctant to give higher tariffs. Therefore, liberal viability gap funding would be required to reduce the tariff which would be fully justified. Hundreds of such plants can be set up.

Three Megawatt Solar PV Plant in Kolar District, Karnataka

The Karnataka Government set up the first 3 MW solar PV plant in Kolar district. Essentially, the idea was to strengthen the rural grid in that area and power electric irrigation pumps during the daytime.[25] This pilot project led to a separate component in phase 1 of the National Solar Mission of 100 MW of 1–2 MW plants feeding power to 11 kVA substations spread in different states. The objective of the scheme was to experiment and set the stage for hundreds of such plants across the country in coming years at suitable places with high day time load. During the year 2010 the plant generated 3.34 million kWh and 3.30 million kWh were sold to the grid (Mitavachan et al., 2011). Evaluation of all such plants at regular intervals would help in developing better system integration practices. It seems that there is currently no thinking of expanding this component in subsequent phases of the Solar Mission.

It is worth repetition that such plants can be very useful in supplying power to the grid at consumption points. There would be reduction in transmission losses and quality of power at the end would improve. The only problem would be the inability of the utility to pay with current rural tariffs. Since these have many positive externalities, it would be worthwhile to provide for liberal viability gap funding, perhaps through interest subsidy, to reduce the tariff.

Cost Competitiveness of Off-grid Renewable Energy Systems

There are several renewable energy technologies, which are suitable for small-scale off-grid power generation. Of these, three major technologies are solar PV, micro hydro, and biomass gasification and a combination in hybrid. There is a plethora of literature on the economic cost of providing electricity from these sources. However, they all provide normative costs. The actual cost of electricity is highly location specific and sensitive to many parameters including distance of the site from the existing grid, road head, availability of renewable energy resource at that location etc. There are additional costs which are discussed in the next chapter because of operating in such areas. The standard procedure for economic assessment is of calculating annualised life cycle cost and unit cost of electricity generation. However, in the off-grid context, the most important deciding factor for going to an alternative renewable energy-based electricity generation project is the expenditure incurred by the rural household in meeting lighting and other productive energy requirement. For example, if a household is spending ₹150 per month in kerosene for hurricane lamps, any option that provides better lighting and less than or equal to the existing expenditure will be welcome. In this background, all the three renewable energy options are cost-competitive. And, needless to say, it is the kerosene subsidy which is the real competition.

Technology

Let us briefly see the technologies themselves.

Biomass technology: Biomass is plant matter such as trees, grasses, agriculture crops, residues and rice husk. Biomass is a renewable source of energy and can be used as a solid fuel, or converted into liquid or gaseous forms, for the production of electric power or heat. Generally off-grid biomass technologies use gasification route wherein biomass is converted (thermo-chemical process) into producer gas in a reactor or biomass gasifier. The producer gas, after undergoing cleaning, can then be used in an internal combustion engine for producing electricity. This method is preferred for small capacity plants (kW range). Biomass gasifier power plants are of two types. The first is dual fuel engine-based where producer gas is used in diesel engine under dual fuel mode (i.e., fuel-mix is a mixture of diesel and producer gas), resulting in up to 60–70 per cent replacement of diesel by producer gas. The smallest capacity of dual fuel system tried in India is 3.75 kW, while the largest plant is of MW capacity, majority of the systems have capacities ranging between 50 and 150 kW. The second one is 100 per cent producer gas engine based, which does not consume diesel for operation and operates solely on producer gas. The standard used for provision of energy access is 32 kW system, which can supply power for about 400–500 households through a mini-grid. Total 10 kW systems for dedicated use for telecom towers are also under pilot test. In such systems, spark-ignited engines (natural gas or modified diesel engine) are used. Various Indian institutions are engaged in the development of this technology. Though thousands of systems can be set up, fuel management will restrict numbers. It also requires good management and technical expertise at local levels.

Micro hydro: Hydro power projects up to station capacity of 25 MW have been termed as Small Hydro Power and accordingly are assigned to MNRE. Small hydro technology is mainly 'run of the river' which operates by diverting part of the river flow or the whole river by constructing a weir and intake, through a penstock (or pipe) and a turbine, which drives a generator to produce electricity. The water then flows back into the river through a civil construction known as the tail race. Recently, small/micro hydro projects are also being built on irrigational canal drops in view of the fact that the irrigational canal flows are already known and the power generated can be quantified. These projects do not envisage construction of reservoirs or dams. Power output of the system may not be determined by controlling the flow of the river in the case of micro and pico hydro projects; instead the turbine operates when there is water flow and the output power is regulated or governed. These provide cheap, independent and continuous power, without degrading the environment.

Micro hydel projects could be actually 'run of the stream' projects and their size could vary from some watts to 100 kW. These are usually suited to hilly areas. There are many in operation in India, though the scope is much more. There are also pico hydel which could be 10 kW or below, including the traditional water mills. Thousand such are in existence, but many thousands more could and should be set up. This is an administrative failure. But location would restrict the total number of such projects.

Solar photovoltaic technologies: SPV systems utilise semiconductor-based materials that directly convert solar energy into electricity. These semiconductors, called solar cells, produce an electrical charge when exposed to sunlight. Solar cells are assembled together to produce solar modules. A group of solar modules connected together to produce the desired power is called a solar array. An SPV-based lighting system typically consists of (a) a solar module consisting of an array of solar cells and (b) balance of system (BOS), consisting of controlling device, battery and support structure and cabling connecting the power system.

There are different ways to use SPV systems for off-grid power generation. Standalone solar PV power plants, basically small power plants, are designed to provide electricity to a small number of households or a village through mini-grids—size and density are important. There could be individual SHS—for interested individuals, whether or not supply of power is there or not or uncertain. It is also suitable for small hamlets and dispersed housing such as in hilly or arid areas. Several models of these solar home systems are available in the market under subsidy and non-subsidy market. There are also smaller solar systems which could power solar lantern charging stations providing lanterns for a few houses or a variety of individual lanterns. This is a growing and developing private market and there is need for better quality control and an organised ecosystem to provide proper services. Solar mini-grids and solar home lighting systems will be the major instrument to bring electricity to India's rural areas.

Small wind generators/wind–PV hybrids: Small wind generators (aero-generators) and wind–PV hybrid systems are used for meeting power requirements in off-grid mode. Depending upon the wind potential of a specific region, different sizes of aero-generators which are in hybridisation mode with PV are installed. The most commonly used aero-generator is of 3.2–3.5 kW range and typically multiples of such aero-generators are installed with different sizes of PV systems. There are, however, hardly any worthwhile field examples of such hybrid models.

We have seen that there are a large number of renewable energy-based energy access projects in operation with different models. It is also evident from the above analysis that there is a business case for deploying renewable energy technologies in a viable manner for decentralised power generation. The numbers can be large, very large. So is the need. What is holding that back is not the cost or the technology but many other challenges and barriers which we turn to next.

Notes

1. Ministry of New and Renewable Energy, Annual Report 2005–06.
2. Ministry of New and Renewable Energy No.15/1/2010-11/RVE dated 26 April 2010.
3. Ministry of New and Renewable Energy. Available at http://www.mnre.gov.in/. Accessed on 22 November 2014.
4. Census 2011 had revealed that around 1.1 million households were meeting lighting energy requirements through solar energy, of which about 0.9 million were in rural areas.
5. Lots of efforts were made in the last 2–3 years of the Plan period which is reflected in the progress made. Letters were repeatedly addressed to Chief Secretaries of states. Regular meetings were held in the Ministry and the Secretary personally reviewed in Assam and J&K (many times) and all the concerned states. Officers of the Ministry made regular field visits.
6. Deepak Gupta initiated this special project and got it approved for around ₹550 crore. It included a package to cover villages and some health and educational institutions with solar PV plants; extensive coverage of solar water heating systems; solar cookers for houses in far flung areas; micro hydel and small hydro projects to provide electrification in remote villages but near water sources; households and commercial green houses to help production of green vegetables and provide livelihoods. Solar PV systems were also provided to army and paramilitary installations encouraging them to go for much larger systems based on this experience to reduce diesel consumption.

Renewable Energy Options for Electricity Access 119

7. The Deloitte evaluation of five states of J&K, Madhya Pradesh, Jharkhand, Odisha and Assam reveals that more than 95 per cent of households across all of the states surveyed are happy with installation of solar PV home lighting systems (HLSs). The rural populace has benefitted with up to 6 hours of electricity, giving rural children an option to study after dark and providing rural women more time for household activities. It also mentioned saving of ₹29 crore as kerosene subsidy.
8. The leader was the Aryavart Gramin Bank at Lucknow which got an Ashden award for this initiative. In Karnataka, this was led by an entrepreneur with initial support from the Syndicate Bank. This story is related in great detail in Miller (2009).
9. Deepak Gupta was Secretary, MNRE that time and surprisingly he first became aware of this lending only through newspaper reports. He then personally visited many villages in 2009 in Uttar Pradesh and Haryana and was greatly influenced by the enthusiasm of the people and their joy with having lighting in the evenings.
10. On 28 February 2014, MNRE extended subsidy for installation of 68,000 numbers of Solar Photovoltaic lights and small capacity systems through NABARD/RRB's, Nationalized banks and Cooperative banks under Jawaharlal Nehru National Solar Mission (JNNSM). Available at http://mnre.gov.in/file-manager/UserFiles/LIghting-Scheme-through-NABARD.pdf. Accessed on 22 November 2014.
11. These details are taken from Miller (2009).
12. Lighting a Billion Lives. Available at http://labl.teriin.org/index.php?option=com_dir&task=all. Accessed on 22 November 2014.
13. At the time of presentation of the budget for the year 2010–11, the government had announced its decision to introduce a special scheme to address the development of 33 LWE affected districts. It was, inter-alia, stated that the Planning Commission would prepare an Integrated Action Plan for the affected areas and that adequate funds would be made available to support the action plan. (*Source*: http://pib.nic.in/newsite/erelease.aspx?relid=79472, accessed on 22 November 2014.)
14. World Bank (2011b) has a detailed examination of the programme and an analysis of the various issues emerging from decentralised energy operations. There is also a detailed technical discussion on the performance of gasifiers and engines.
15. GIZ. Available at http://www.giz.de/Themen/en/30546.htm and discussions with GIZ. Accessed on 22 November 2014.
16. Indo-Norwegian Pilot 'The 30 Village Project' Key Learning's—2nd July 2012. Available at http://haritika.org.in/Presentation,solar%20%281%29.pdf. Accessed on 22 November 2014.
17. Financial assistance under Remote Village Electrification scheme 2009–10. Available at http://www.mnre.gov.in/file-manager/offgrid-remote-village-programme/rve-adm-2011-12.pdf. Accessed on 22 November 2014.
18. This information is based on discussions with officials of the company.
19. The case study draws from various sources, including WEBREDA, Sam Shrank, Working Paper #77 'Another Look at Renewables on India's Sagar Island', The Program on Energy and Sustainable Development, Stanford University, and Rural Energy Initiative's Best Practices in Sustainable Rural Energy Development.
20. Presentation made by HPS in the Green Opportunities in Tomorrow's Markets: New Ventures Global Investor Forum on 6 April 2011. Available at http://www.new-ventures.org/company/husk-power-systems. Accessed on 22 November 2014.
21. Access to Energy: A Glimpse of Off-grid Projection India. An MNRE publication.
22. It is understood that the Solar Energy Corporation of India (SECI) has initiated a procurement process to distribute 6 lakh small solar lanterns.

23. Biomass Gasification Based Power Generation by Arashi Hi-Tech Bio-Power Private Limited; CDM project monitoring report. Available at http://cdm.unfccc.int/filestorage/6/D/4/6D4CG7M5BJFYV0PO1KZI8WHSX2NRLQ/MR_Arashi_V4_05102010.pdf?t=VFV8bXZtbW5sfDABeiHUEdhm_QLuwSs1813Q. Accessed on 22 November 2014.
24. Idea for this plant was initiated following Deepak Gupta's visit to the area when he was Secretary, MNRE and discussions with forest officials about how to find productive avenues for pine needles.
25. Deepak Gupta's visit to the plant led to the incorporation of such plants as one component of the first phase of National Solar Mission.

6
Challenges for Universal Electricity Access and Way Forward

Background

The review and discussion of the power sector problems in Chapter 3 and the RGGVY programme in Chapter 4 lead to three conclusions. First, household electrification is less than satisfactory in large areas of the country. Second, the centralised grid approach is unlikely to resolve this problem substantially. There will be some areas where electricity supply will indeed become better. These could be areas that are nearer to towns or more developed habitats or along major roads and where markets have been established. This will leave out the hinterland, the remote areas and small hamlets almost everywhere. Therefore, in thousands of villages/hamlets the situation will remain much as it is today, with uncertain and unreliable power which is largely absent when needed most in the evening hours. This leads us to the third conclusion that renewable energy solutions will be necessary. The evening hours certainly involve essentially lighting but there are also other opportunities and possibilities. There could be household entertainment through TV and a more vibrant life in the village with street lights and shops remaining open till later. There can also be increased income generating activities within households and even some micro-industrial activity during those hours when small machines could work to serve local needs. The RGGVY evaluations have tried to list out benefits of rural electrification. For a typical village with normal demand, it is clear that these benefits could also accrue from power supplied by renewable energy systems. Indeed, in many areas/villages these are currently being supplied through diesel-based micro-systems. The potential is unlimited because there is no real finite barrier with fuel constraints not being there. This potential now needs to be converted into the reality of reaching to about 300 million people.

Electricity for the rural areas has generally and traditionally been understood as extension of the centralised grid to the villages. In a sense there really was no alternative earlier. Even in the many villages provided with lighting under the RVE programme, people felt that they had been served with second class electricity. Such electrification was, therefore, considered an interim solution as ultimately the grid was expected to reach these regions. Therefore, irrespective of the uncertainty of power, load shedding, and rostering schedules that dampens demand for, and availability of, power in rural areas, there cannot be any doubt that grid connection remains the most favoured approach to rural electrification for the majority of rural households. The preference for grid power, however, goes well beyond the convenience factor. One of the legacies of past rural electrification projects has been first to deliver electricity as a free or highly subsidised good and then subsequently fail to implement effective charging schemes to secure the continuation of power supply by ensuring revenue sustainability. In many rural areas, the supply may be so dysfunctional, and the tariff collected relatively so small, that the cost of billing/collection may not be worth the effort and cost. Here electricity will be virtually free. There will also be rampant theft and bill default. The inability of utilities to supply regular, reliable and quality power has only strengthened this attitude or behaviour. The perception then is that it is cheap or free. This has crept into the psyche of the people. We have created a feeling that this is a government dole. People have accepted the much higher payment for kerosene because there is no choice. Alternative solutions have both to give a feeling of electricity while creating a climate of paying for service that implies charging a proper amount for renewable generation and getting regular payment for it. Both these would be challenges. Had the kerosene subsidy been less, the task of replacement would have been much easier, as will be seen in countries where there is no such subsidy. But this approach to kerosene subsidy in the context of rural electrification has never been considered because kerosene is seen as a poor man's fuel.

Access programmes have also been driven by target-oriented supply push strategy and subsidy-driven national programmes. This has resulted in greater emphasis on meeting the physical targets rather than on effectiveness, and even sustainability. The existing regime does not incentivise cost reduction as subsidies are linked to capital cost and are measured in terms of percentages. Since successful electricity access programmes would be contingent on widespread willingness to pay (WTP) amongst rural households and energy users, a more robust and sustainable solution is necessary, which would come from entrepreneur-driven business models. The task of the government, therefore, needs to be an effective and efficient facilitator, helping develop the larger as well as the local ecosystem and provide the financial wherewithal. Subsidies and debt would be required. The question then is: what is the best possible way to do this and ensure long-term sustainability? Cost recovery for the developer is probably the single most important factor for this. When cost recovery is pursued, most of the other elements fall into place (Barnes and Foley, 2004). This then becomes the fundamental challenge.

In this context, it must be recognised that low village demand coming from low population densities in rural areas results in high capital and operational costs while revenue collection is poor and at high cost. To an extent, the challenges faced by the utilities in supplying power to rural areas will be there for local supplies too. But localised solutions from renewable energy sources by providing generation at consumption point could convert these low levels of demand into an opportunity and make them viable.

In the earlier years, cost of solar was perceived to be very high. Even then it was recognised that solar was cost competitive in rural un-electrified markets in meeting the limited needs of the people. However, the problem clearly was, as it continues to be today in spite of reduction in costs, that the upfront costs were very high because one would need to pay for a major part of life time costs at the start itself. And this cost was high relative to annual income. This led to the conclusion that 'lack of relatively large cash sums required for the purchase of PV installations is the most significant barrier to their wider use by rural families' (Foley, 1995). If cost alone was the issue then the dramatic reduction should have immediately led to widespread adoption. Therefore, we need to analyse in depth and consider all the factors which inhibit large-scale dissemination. If these barriers remain, even further cost reduction would not lead to increased spread. It also implies that policy should not wait for further reduction in costs to increase the adoption.

We have also discussed and reviewed in Chapter 5 the various efforts made initially by the government and over time by others including NGO's and entrepreneurs for providing solutions through use of renewable energy. Various models have been tried and are being attempted. We can also call them pilot projects as none have reached scale, except the RVE scheme, which was almost fully subsidised and whose sustainability is open to question. There have been almost a hundred projects based on husk power and several solar mini-grids. All these projects have together thrown up various issues and identified barriers and important challenges that include credit constraints, proper funding support, lack of technical capacity, lack of awareness, underdeveloped market conditions, willingness/ability to pay, etc. These challenges are altogether different and unique for off-grid projects except, to a certain degree, the issue of tariff. We now discuss the barriers in order to find generic solutions to overcome them and then suggest a way forward to achieve scale which may eventually lead to genuine universal electricity access in India. This will require alignment of economic incentives, development of sustainable market conditions, facilitating institutional structures and the push of determined state policy.

Sociological or Organisational Challenges

These are the special challenges for small isolated systems to be set up and managed on a commercial basis in rural areas. These start, as one of the entrepreneurs told us, with the locational and logistical problems such as finding contractors, supplying material, lodging of engineers and doing the installations. Entrepreneurs also need to invest in establishing network of sales staff and technicians and their training, vehicles, etc. These may appear to be simple issues but could cause huge problems, require a lot of effort and add to costs.

Other organisational or sociological issues arise from the rural and village context where the plant will be set up. Will the community welcome this endeavour and be reasonably united? Entrepreneurs working in the field have reported community mobilisation as an issue, apart from this taking a lot of time, and adding to the cost. There is a view that people feel deceived by past promises of delivering electricity leading to the problem of building of trust by a new entrant in the area. This trust is essential for starting a mini-grid and will be necessary in successfully running it.

There are payment issues. Will everybody pay for the initial connection fees, which would be essential as a security and to get the commitment, and make regular payments thereafter? Will there be default and whether that will be ignored so that it does not spread? Both Husk Power and Gram Power have developed prepaid systems. Collection of small amounts from many is a problem, though not as big as it is for the utilities. So, experiments on collection methods have to be done. Harish Hande of Solar Electric Light Company (SELCO) has been propagating the concept of daily payments as it allows everybody, even daily wage earners, to pay as they earn. This should be possible in a mini-grid. Will people use as much as they are metered for? Will they tolerate some local problems? Will they provide easily the small amount of land required? There should be an assurance of safety. Will one of them become the local technician and give proper service? All these issues would call for a comprehensive awareness generation programme helped by the government. At present, this challenge is being met by the entrepreneurs themselves. Needless to say, a local entrepreneur would be much better placed to manage these issues unless there is serious village politics and divisions.

Entrepreneurial Challenges

There are challenges related to which organisations would want to or could set up projects. This is fundamental. Will it be an NGO or a large company or an individual entrepreneur, whether from the local areas or from outside? Let us briefly discuss.

NGO

An NGO has good reach and local capacity but its capacity is limited to its area of operation and it cannot upscale. Generally they are not in the business of setting up and managing projects. That is why they have largely been involved with pilot projects only. Grameen Shakti and BRAC in Bangladesh, despite being NGOs, operate more as businesses.

Local Entrepreneurship

Individual entrepreneurship is currently a critical constraint in rural areas and may continue to be especially for somewhat technical and difficult projects. It is not that entrepreneurship is lacking. There are already many who are operating diesel-based systems. It would be necessary to upgrade their skills and confidence to both enable and motivate them to operate renewable energy systems. In our view this may become essential for large scale-up. Can the women self-help groups become the entrepreneur? This we feel would be the most ideal solution. So far there is no such example but it would be desirable to try a pilot through Corporate Social Responsibility (CSR). Can this area also excite some rural youth who have some funds or access to them and some enterprise and are willing to remain primarily in rural areas? Local entrepreneurship could be the key, quite apart from launching an entrepreneurial revolution.

Big Business

Larger companies have generally shown little interest, although they have the capability. Many small projects spread over an area do not suit their business style and rural areas appear to be far away from their comfort zone, especially those where the requirements are the greatest. They usually do not see big opportunities or large profits in rural areas especially compared to the efforts involved. They may bring in strong managerial capabilities but their process driven systems would lead to higher costs and inability to take quick decisions, apart from their strong aversion to risk. A rural energy business would only be part of a diversified revenue stream and they may not have staying power or the will to do so. Shell Solar did well in Sri Lanka for 6 years but then exited (Miller, 2009). British Petroleum (BP) entered the cookstove area in 2008 and then exited shortly thereafter. So did Philips. It is doubtful, therefore, if large professional companies could be persuaded or incentivised to come here in a big way, nor probably would that be the best way forward. However, the World Bank is currently proposing an Energy Access project in India hoping that that will happen.

Some companies have come forward. Bosch energy is working with Gram Oorja. Their 10 kW system in Darewadi is being run by a Trust. Sun Edison has been working in many villages through their Foundation. They are also trying to have some projects under the DDG scheme of the REC. Schneider is proposing to cover many villages. This area is ideal for CSR and could help develop new local entrepreneurs. Minda Nextgentech Ltd. has chosen this route. CSR could help set them up by meeting larger initial costs and then through the experience and learning they could spread further on a more sustainable basis.

Other Entrepreneurs

The most exciting and promising are the young entrepreneurs who have come forward and have set up the several projects that we hear about. We can call them the first movers who are showing above average dedication and leadership. That is the real value of Harish Hande, Gyanesh Pandey, Yashraj Khaitan, Damien Miller, Sameer Nair and others. Many more are needed. Such entrepreneurs have come of their own will driven by a special desire to work in this area and with almost single-minded determination to pursue and succeed. Their biomass- and solar-based entrepreneurial efforts for local decentralised supplies are really the first examples of an effort to look at the rural electricity deficit as a business opportunity. Each of them has struggled to ensure viability of systems at an affordable cost while making a profit. One cannot say that they have been fully successful as full details of the source of and cost of funds are not available nor of the incomes earned. Sources have been many including venture capital, private investments, social capital and a mix of government subsidies. A clear picture and a model have not emerged. They have tried to reduce costs, ensure efficiency, mitigate risks and provide reliability. These are the people we need to welcome with open arms and more than incentivise. But do they have the capacity and the financial wherewithal to grow exponentially? Can policy provide both? That is our challenge. We are confident that all this will happen and there will be different models with different entrepreneurs. All this demands that the state respond positively and unambiguously to the issues thrown up and support their efforts so that this business opportunity could become big enough, not in dollars earned, but in size to attract hundreds of people like them. It will lie in their hands to solve India's problem of electricity access. Perhaps the answer will lie in a combination of local entrepreneurs, including women self-help groups, who will invest in and operate the systems. Some experiments are already there.

Technical Challenges

Technology choice: Several years ago, in an important study on rural electrification challenges for renewable energy based on analysis of several projects (Cust et al., 2007), the authors had suggested that the most significant initial barrier was technology choice. This was perhaps because they had studied many village energy security programmes (VESPs) and other projects based on biomass gasifiers as solar was very expensive at that time. Technology should not be a real issue now. In fact, the challenge is to dispel the myth that renewable energy technology (RET) will be unreliable or expensive. It is true that biomass gasifiers have issues—working of the engine and the gasifier itself, which also require competent technical people and regular maintenance (Palit and Chaurey, 2011). Several local innovations have been done which have reduced costs and improved performance. We have mentioned earlier that all the different technologies are now mature and cost competitive.

Balance of systems: Perhaps the important areas for technical improvement are the balance of systems—charge controllers, small inverters, batteries, prepaid meters, etc. We have noted several problems—poor-quality inverters causing inefficiencies and breakdowns; replacement of broken CFLs by ordinary bulbs increasing power demand; difficulties in replacement of CFLs, batteries, charge controllers, etc.; difficulties in fixing problems because of lack of spares and skilled personnel; battery management, etc. It would be useful to have a comprehensive technical study done in these areas. This would lead to lessons learnt so that remedial measures can be taken.

Fuel management: In the study mentioned earlier, the authors had pointed out that fuel management was also a significant challenge. This conclusion arose from the difficulties experienced in the VESP projects where fuel supply was a community responsibility. But for rice husk or corn

Challenges for Universal Electricity Access and Way Forward 129

cobs or others that are used in projects in Bihar, this has not been a major problem because the systems are small and fuel is locally available, which will continue in the future too. There would, of course, be increased managerial responsibilities as gasifiers impose a day to day requirement for procurement, storage, payment and use of fuel. However, there would be a limitation in the number of projects that can be undertaken and may define the areas where this is possible to be done in a sustainable way. A large number of villages in Bihar and east Uttar Pradesh should not have a problem. There could be problems if mini-grid sizes are of megawatt scale as is being discussed in the proposed World Bank Project One could imagine a somewhat remote area in northern Bihar from where paddy is currently being taken to towns for milling. Maybe a local entrepreneur could set up a small rice mill powered by a gasifier in dual mode and provide the rice husk for village-level gasifiers. Or there could be many such. Wherever dedicated plantations are grown fuel supply would not be a problem. Bamboo in the NE provides many possibilities. These systems, therefore, would require more technical and managerial skills and more experienced entrepreneurs than the solar mini-grids.[1]

Technical manpower: The other major issue is availability of technical manpower. Solar systems are still technology intensive and there is a mystery about them. It can be 'intimidating' for villagers unfamiliar with technology (Deorah and Leena Chandran-Wadia, 2013). Even though maintenance of solar systems is not a real problem, there is need for technical support and management. Skilled personnel must be local. The challenge then is to locate such local people and train them to be willing and able local technicians. This is necessary for proper provision of service, which in turn, is necessary to make solar an integral part of rural electricity development.

Adequacy of load: We have seen how low load levels affect the economics of grid supply. The question raised is whether provision for lighting, that too for a few hours, would make a local system viable? This is a challenge. However, studies show that there is WTP enough for light only to make even a small solar project sustainable over the long term. As we noted in the Chhattisgarh mini-grid review in the previous chapter, demand of many villagers increased. This is a sign of acceptance.

Capacity of the system can be increased through modular upgradation of the solar plant. It is difficult in the gasifier case. There are also reports of households taking loans for larger systems under the bank-financed solar home system (SHS). Loads for shops and micro-industries can also be managed. These can be charged at higher levels because the WTP for commercial operations should be more. Meeting such small loads would add to the viability and sustainability of the systems. After all a normal village does not have more loads than that. Adequacy of load, therefore, while continuing to be a serious problem for grid supply, should not be a real problem for local decentralised systems. A cluster of such systems in nearby villages would make the projects even more viable as fixed costs can be shared and total revenues increased. But returns cannot be expected to be very high.

Anchor load: In the context of load adequacy and revenue sustainability, there has been talk of an anchor load which will provide higher revenues. Desi Power has been working on this concept for many years in Bihar but has found it enormously difficult to set up such systems. It is not easy to find anchor loads.

Supplying power to both the telecom tower, which uses substantial quantity of diesel at great cost, and the village is not possible through the same renewable energy systems, whether solar or gasifier, because the other loads are only for a few hours while telecom towers require power for 24 hours. Some experiments are being tried. A dedicated small gasifier has been developed for telecom towers only and a few are now being reportedly funded by IREDA.[2] Omni-grid Micropower Company (OMC) is an example of a successful model to use telecom towers as anchor load, though details of their costing and recoveries are not very clear at present. Therefore, it would be useful to work for the evening hours only as this is the prime requirement in thousands of villages. But there are future possibilities. The entrepreneur who manages a mini-grid could provide a separate line to the telecom tower, maybe agreeing in the first instance to provide power in the day time, or provide to any other daytime load which may develop in time such as for a mini enterprise. Looms in village in Jharkhand are an example. It is possible for an entrepreneur to install extra solar capacity to charge lanterns and mobile phones. So far there is no such example. These possibilities will develop in due course.

Regulatory Barriers

Under Section 14 of the Electricity Act, 2003, any person who wants to generate and distribute electricity in a rural area notified by the government need not require any license. Pursuant to the National Rural Electrification policy all the states have already notified rural areas for the purposes.[3] However, the situation is ambiguous and liable for different interpretations. This was the problem faced by Desi Power in their original project and they had to adopt a very circuitous route by involving a cooperative in purchase and distribution. After Husk Power had set up their systems in Bihar, Bihar Renewable Energy Development Agency (BREDA), the Agency set up to promote renewable power, had raised serious objections and threatened cancellation because of this issue. It was only after the intervention of the MNRE with the state government that the problem was resolved, and only after a couple of years that the government of Bihar notified rural areas where licences would be exempted. Clarity needs to be ensured in all states. There may be other possible emerging regulatory threats which are discussed below.

Issues Related to Tariff, Costs and Subsidy Support

This has three aspects—what is the cost of supply to the generator, what is the ability or WTP of the consumer, and keeping these in view, what support should be extended to the entrepreneur. Together, they constitute the crux of the challenge.

Tariff issues: A study has advocated that off-grid power should also be regulated and that the tariff payable should be determined by the Regulator.[4] It has proposed a model in which the project developer shall provide electricity to the consumers and collect the tariff as paid by the consumers of the local Discom. The Discom shall provide the balance feed-in tariff to the project developer. The Government of India would provide CFA to the Discom for promoting off-grid rural electrification.

In part this model was adopted in the draft guidelines issued by the MNRE in March 2012.[5] In a major shift, both understandable and desirable, the programme was sought to be made applicable to existing electrified villages. However, this applied to villages where power was received for less than 6 hours a day. This is vague because those 6 hours are not defined. A certificate from the state power department was proposed. The question is who will seek, and who will give, such a certificate? We have already suggested a changed definition for an electrified village. If that is adopted such villages would be automatically listed. Street lights were to be mandatorily provided. The question of who would pay the tariff for them has been left unanswered. Tariff payable by the household would be fixed by the State Agency. The project developer would handle the projects for 5 years and then transfer to a state designated agency, like in the present scheme of the DDG. So sustainability is not built in. State governments would replace batteries, an undefined future cost and action which the states may not be able or willing to do. The subsidy levels were kept at the unacceptable level of 90 per cent with subsidy now to be released in five instalments. Imagine the task of doling this out if 10,000 villages were covered, or of developers trying to access these? Clearly, therefore, the idea was well intentioned, but impractical and unsustainable from the start—state-led and regulated; immensely high subsidy driven; and with no long-term incentive or desire for the developer. It is no surprise that there have been no further developments on this front. Basically, this was also meant to be a grid interactive scheme.

Theoretically, the argument for providing a feed in tariff is valid. Consumers in rural areas would pay rates at par with their urban counterparts for grid charges. There would also be certainty of revenue for the developer. However, we see two major difficulties. First, bringing the Regulator, or the State Agency, into this area is fraught with some risks and possible difficulties. There is a general impression that the Regulators have not been able to do enough for grid-based renewable electricity, including tariff setting for conventional power. This will be a new and complicated area with many uncertainties and there could possibly emerge a maze of regulations which could more likely constrain both innovation and enterprise. One worries what could be the tariff set for rural areas. As far as State Agencies are concerned, they have neither the authority nor competency. Second, the power generators would then be expected to go to the government or the state utility to get the balance of the tariff. As it is capital subsidies are hard to access. A tariff concession will most likely face much greater problems. Imagine a situation if the utility does not get money from the government, or not in time, and in turn, delays payments to the developer! The basic tenets of off-grid are off-regulator and off-utility too. At present we need to keep this sector free from the Regulator, the State Agency and dependence on the utility with all its problems. The DDG scheme is hardly working even with an abnormal amount as subsidy. There are currently only a few mini-grid projects where people are struggling to find sustainable solutions because of lack of proper financial support. It is not that hundreds of entrepreneurs are rushing to get hold of a potential pot of gold—in fact, finding entrepreneurs is one of the biggest constraints in scaling up. After many hundreds of such systems are in place and run for a few years we can think of such a tariff. By that time kerosene costs should be higher and so should the urban tariff. The real case is actually for subsidy reform of kerosene and revision of urban power tariff. For the present, we should allow systems to be set up on the basis of the willingness of consumers to pay but find the needed financial support to be provided in a satisfactory way to ensure tariffs set by developers are not above the WTP while providing for a decent return to the developer. Independence, autonomy and sustainability are the key words.

The amount of tariff is also difficult to estimate as it would depend upon incomes, existing energy sources and their cost, availability of electricity and quality of supply. There is a lack of quantity and quality of data. Different numbers have been quoted over the years by different surveys and studies. Essentially, there is WTP kerosene replacement cost. This may also include the cost of accessing kerosene, and sometimes even paying the market rate for part of the purchases. There is also a need to factor in the cost of charging mobiles. But various studies generally support findings that households are likely to pay more than the existing subsidised tariffs. Even where there are existing grid connections rural households often pay effectively high tariffs when the costs of connection, fixed charges or monthly minimum charges are included. It would be desirable, therefore, if there is no fixed tariff and we leave it to an arrangement between the supplier and the receiver. This was also the intention in the Electricity Act 2003 which made under Section 8 the exempted licensee generating and distributing electricity in a rural area free from the purview of the Regulatory Commissions in matters pertaining to determination of tariffs.

It has been argued that it would be equitable for rural consumers to pay existing grid tariffs. As far as equity is concerned, the principle would be better served if supply of power is ensured rather than people remaining dependent upon high-cost kerosene for poor quality lighting. In fact, this principle would strongly suggest a review of urban tariffs (as also costs for cooking gas) as well. If over 300 million people are currently paying the cost of kerosene for getting minimal light, it would appear rational that the minimum tariff for all should be at levels paid by these users. Indeed, if that were done, most of the problems of the power sector would disappear.

Willingness to pay: In this context let us discuss the issue of the WTP. Studies over years and different areas suggest electricity consumption has high value for rural households, and where access exists, WTP is high, even amongst poorer households (World Bank, 2004). These studies conclude that facts do not support the widespread belief that electricity tariffs need to be extremely low, often well below their cost of supply, to benefit rural people. In the absence of grid supply, demand for electricity using services is met by spending money on kerosene, batteries or diesel, all highly expensive per unit supplied. RE tariffs set at realistic levels do not prevent people from making significant savings in their energy costs as well as obtaining a vastly improved service. Consumers are generally willing to pay significantly more for shorter outages and better quality supply even in grid connected areas. Therefore, the issue should not be the level of tariff but the proper supply of electricity (Cust et al., 2007). Anecdotal evidence and surveys suggest that households, depending upon the amount of lighting they require, have been spending between ₹50 and ₹150 per month on fuel or more. All the renewable energy technologies, with adequate support, can supply power at this rate and that of the diesel cost. Should the subsidy on kerosene and diesel be respectively reduced or eliminated, the shift to renewables could become that much easier in rural areas. In Africa and our neighbouring countries, kerosene is not subsidised. Therefore, it would be much easier to shift to renewables.

Challenges for Universal Electricity Access and Way Forward   135

Cost issues: State utilities typically report an average cost of supply at around ₹3–5/kWh of grid-based electricity. However, an estimate for a Gujarat case study in 2007, based on Gujarat Electricity Board data, put the true cost of delivery to rural areas at over ₹9/kWh. It is also suggested that supply costs increase by roughly ₹1/kWh per km of expansion to individual villages (Cust et al., 2007). The World Bank study of 2010, referred to in Chapter 5, also suggests that the cost of electricity in rural areas is somewhere around ₹11/kWh. While numbers may vary, it would be reasonable to assume that the actual cost of supply would be much more than the tariff. We have also computed the cost of evening hours supply in Chapter 3.

Supply through solar in rural areas has some additional costs. The spread out low population, along with the low load, increase both the total and the per capita cost of electrification. The initial cost of installation becomes higher also because of the cost of transportation and manpower and subsequent maintenance. There are other factors also in increasing cost. While solar panels may have become cheaper, battery costs have not and may have probably increased. And substantial capacity of batteries would be needed. The use of battery means that there will be replacement costs which have to be considered. Further, there are additional per unit costs because unlike in a grid system, all power does not get evacuated. A separate local grid also has to be constructed. There will also be differences in costs depending upon the size of systems, with smaller ones being more expensive. The systems that use three days of battery autonomy would be quite expensive. In fact we would need to consider what should the battery autonomy period ideally be. Perhaps there is a case for simply one day with adjustments being made for the amount of sunlight available on particular days as far as the consumer is concerned. There is certainly a strong case to fix differentiated battery autonomy criteria in the country taking into consideration the general climate pattern. Normal norms of grid or urban off-grid rooftop, therefore, will not apply. These are important issues for determining the level of subsidy support.

Subsidy issues: Let us now consider the two problems with existing subsidy, which is being given by the MNRE to facilitate setting up of mini-grid systems.

Extent of subsidy: The MNRE has been giving a subsidy of 30 per cent of the benchmark cost. The assumed cost for mini-grids was ₹500/Wp. The support was ₹150/Wp. This included having a 5-year O&M contract and other guarantees. There was additional support for setting up the distribution infrastructure. However, these benchmark costs were drastically reduced in May 2013 to 350/Wp for up to 10 kW mini-grids and 300/Wp for above 10 kW capacity.[6] These reductions appear to have been made in view of the general price reduction in the cost of solar PV cells. However, there is a general feeling that this level of subsidy is unrealistic with a system for three-day battery autonomy and may hurt the fledgling efforts being made by some entrepreneurs in this area. On the one hand, we want to restrict tariff. On the other, subsidy is sought to be reduced. Perhaps there has not been a realistic evaluation of costs. Compare this with the hidden and explicit subsidies that continue to be given to fossil fuels and instruments like generation-based incentive (GBI) and viability gap funding (VGF) for grid renewable power. It is a fundamental contradiction that the DDG scheme of the MOP gets 90 per cent subsidy against an unspecified benchmark cost for a project which should last 5 years, and that a similar scheme under the MNRE gets 30 per cent of a reduced benchmark cost, where the project has to remain sustainable for a much longer period and will also involve periodic change of batteries. It is no surprise that the MNRE draft guidelines of 2012 were also suggesting this high level of subsidy, even as the 30 per cent continues to be retained in the existing scheme.[7] REC has supported 276 projects in seven states for 5,624 kW at a cost of ₹180 crore.[8] Clearly, therefore, one is too much and the other is too less. A fixed subsidy is better and ₹140–180 per watt for different sizes prima facie appears to be the right band. In addition, there should be a 90 per cent subsidy for the distribution infrastructure to be set up. It is necessary that the government come out with a detailed analysis, much like the CERC does when setting a tariff, of how this subsidy has been calculated. A liberal subsidy will replace a much more liberal and damaging kerosene subsidy. This should be decided by a consultative process while looking at the real costs and the time period for a project which would then also consider a realistic period of battery autonomy and battery replacement costs after 5 years and the implications thereof. It will also replace a much more liberal DDG scheme. It is essential that we incentivise such projects to encourage entrepreneurs to set them up in the difficult conditions of rural areas. Subsidy amounts can be reduced for solar rooftops for urban areas with time. But for rural areas such a subsidy level should be retained for a period of 5 years, to have consistency and enable people to plan without worry of uncertainty and sudden changes, which unfortunately keep happening. Solar modules costs may come down marginally but all other costs are not. Therefore, while there may be a case for its gradual reduction after a few years we may not be able to eliminate it altogether.

Procedure to get subsidy: It is also important that the procedure for the disbursement of subsidy is also streamlined so that accessing it does not become too cumbersome. In the earlier years, this subsidy was given in three instalments, which increased uncertainty and transaction costs tremendously. After a feedback from the entrepreneurs, this was changed to a back ended subsidy. The disadvantage was that funds were required upfront for the entire investment and subsidy disbursement being uncertain, especially in terms of time; it put a huge financial burden on the developer. There was also no instrument to manage the bridging period. But now almost all entrepreneurs are complaining that this procedure is also becoming too cumbersome and difficult. First, the project has to be approved which takes several months. Once the work is done, there is a field inspection. A report follows where after the money is released, which takes many months. There are entrepreneurs who have reported pendencies. What happens to the loan liability and risk, and even the conviction, of the entrepreneur, who has made brave efforts to venture into this difficult area if delays occur? Delays could drive away the first-time investors. We must find more robust procedures.

Grid Interactivity

Let us consider another question which is repeatedly raised—what happens to the mini-grid system when the grid comes? The answer given is that the mini-grid system should be such that it could be integrated with the grid when that happens. Perhaps the World Bank proposed model is also keeping this in mind. The suggested models above and the MNRE draft guidelines also emphasise this. But this thinking ignores more relevant ground realities. Most places where such systems are being currently set up, and will in the future, already have the grid. These are not the remote areas. So the real question is what happens when the power supply improves? Clearly, if power supply improves, especially in the evening peak hours, and the tariff is kept low, people will shift and the renewable energy system will collapse. But we are talking of areas where such supply of power is very unlikely to come for years—well beyond the working life of the system. Such systems should first be set up in such areas which would be identified by the mapping. After all, the mini-grid is not going everywhere. So there need not be any worry at all on this score.

The Observer Research Foundation (ORF) report even hints at a possible bias in the thinking at higher levels (Deorah and Chandran-Wadia, 2013). They point to the repeated discussions mentioned earlier regarding regulation of tariff and the focus on grid projects. They also point to the Annual Report of 2011–12 of the MNRE which states: 'The National Rural Electrification Policy, 2006, has clarified that provision of renewable energy-based systems in un-electrified villages and hamlets should not jeopardise the rights of such villagers to grid connectivity'. Since we believe that the grid is still far away from the remaining such villages, and more importantly, the grid will not bring power, therefore, as Carl Pope of the Sierra Club says quite rightly:

> It is unconscionable for those without energy to be forced to continue to spend huge portions of their monthly income on dirty kerosene and diesel instead of cheaper and cleaner solar and biomass because we want them to wait for grid power to arrive. A grid that (really) hasn't come for decades, and will not come for decades more. (Deorah and Chandran-Wadia, 2013)

We have over 300 million people without energy access. Let us target at least half through this way and we will address this question then. Where it is expected that power supply may improve, the bank-financed individual solar home lighting systems will work. In a large measure, the mini-grids will be independent grid systems for a village. They are producing small amounts of power, which will not be able to supply to a partly functional grid. They are also supplying essentially for a limited period of time, even though models may be developed which supply for longer hours. They have also to be designed so that their full cost is recovered in a few years, after which there is only a maintenance cost, subject to battery replacement. At that time the utility concerned can use the system by paying a management cost to the generator and letting that power be continued to be distributed in the peak period or buy the depreciated assets and use them themselves or pay a tariff and have the power fed into the local grid. But let us discuss this issue after a few years. Let us first do much more work as also develop an ecosystem. Asked this question, Sameer Nair of Gram Oorja responded thus:

> Decentralised generation with an ability to interact with a grid could in theory be a useful solution. However, in practice without clarity on how grid interactivity can happen technically and without a clear understanding on commercial terms of supply (like net metering etc) this will not take off. (Deorah and Chandran-Wadia, 2013, p. 43)

Let us not, therefore, raise questions regarding problems which may not occur right at the very beginning. We will simply lose our way by first trying to look for such solutions. There is, therefore, no defined exit strategy for the moment and there need not be.

Access to Capital

This brings us to the key financial sustainability issue, which is access to capital. Ideally, from a business model point of view, there should be a private investor, either the entrepreneur himself or as a promoter. He should be able to access finance easily. But from where will this come? It has been the general experience of the last few years that the expectations of private financiers have been unusually high. High returns are simply not possible, and the perceived risks are also high. For the present, therefore, this source seems unlikely as a sustainable funding source. This area is actually thirsting for social investors, of which there are many in the world, who will accept much lower returns because they are getting also the satisfaction of improving the life of the poor. More of these need to be tapped.

But the real need is for the banking sector to come forward in a big way. The banking community has a general apathy to fund decentralised renewable energy-based electricity access projects due to their old perceptions about cost-effectiveness and the financial viability of such projects. IREDA, a key financial institution in the renewable power sector under the MNRE, unfortunately does not support such renewable energy projects in rural areas, and is reluctant to start. Small Industries Development Bank of India (SIDBI), set up to promote small enterprises, has not come forward. Power Finance Corporation (PFC) lends to major generation and transmission projects and grid projects. REC is lending to renewable energy grid projects, but, in spite of its dominant role in rural electrification, does not concern itself with such projects even though it is also the coordinating agency for MOP DDG projects, which are dependent upon huge subsidies. Rural banks have similarly not come forward except to promote some solar home lighting as discussed separately. It is a fact that the entrepreneurs would be mainly from neighbourhood locations who would have limited financial capabilities and also entrepreneurship abilities. The risk perception, therefore, becomes greater. There are other impediments—unfamiliarity of the local and higher level banking community; the need for collateral from local entrepreneurs and the lack of confidence in them, in the absence of both their credit history and past performance.[9] The risk-averse nature of the commercial banking sector is reinforced by a lack of demonstrated medium term success that could build confidence and capacity for lending.[10] It has been said, somewhat uncharitably but largely true, that 'capital is often made available by reluctant banks to English speaking entrepreneurs with a pedigree but vernacular entrepreneurs who (will) drive scale will find it hard to access capital' (Deorah and Chandran-Wadia, 2013).

But access to location-specific cheap financing is a prerequisite for a successful electricity access programme. Therefore, the bankers have to be brought on board and convinced that this funding is a part of a national programme with its priorities. With a 50 per cent subsidy, half of the loan would in any case be recovered. Poor recovery in agriculture sector which has been about 75 per cent has worried lenders. Part of the problem has been created by loan waivers. This further tends to accentuate risk perceptions for rural bankers. But energy access is different from crop lending. Further, lending under bank projects for home lighting systems have generally found more than satisfactory pay back. The Finance Minister, while speaking at the anniversary celebrations of the J&K Bank in Srinagar in September 2013, made a passionate plea for lending to the poor. Banks need to translate this passion into real and

Challenges for Universal Electricity Access and Way Forward 141

facilitating instruments and attitudes. In this context, the training, which would actually also be a sensitising process, of banking personnel will play an important role. But there is also a need for revisiting bank procedures for such projects. Loan criteria and conditions, including collateral and credit history need to be re-examined. Some other smaller things may have to be required such as no need to get approvals for land use for the land required for setting up of the systems, which are hardly much. The off-grid projects would require customised finance. So, financial products are required to be constructed.

The RBI has included loans sanctioned by banks directly to individuals for setting up off-grid solar and other off-grid renewable energy solutions for households under priority sector lending.[11] This is a very welcome step. It is important to operationalise the priority sector lending provisions for the distributed energy projects. They should not be monopolised by the large conventional power sector, which is already facing strain. A separate dedicated window needs to be created for such projects. There was a proposal from the MNRE that in case the banks fall short of their obligation for lending to off-grid renewable, the resources of banks would need to tap into to the extent that they fall short of their obligation to lend to the priority sector, by setting up a Renewable Energy Development Fund (REDF). However, the RBI did not accept the proposal. The Trust Fund could now serve this purpose. But there is also a need to set up a suitable low cost risk guarantee fund especially established for this lending which would provide both the confidence and actual support to cover the risk perception of banks.

Another segment within the Bank Finance relates to availability of microfinance and involvement of women in such operations. There are shining examples, not only in India but in other countries too. In many of the business models being tried, like SELCO on a small scale in India and Grameen Shakti in a much bigger way in Bangladesh, users are willing to pay as long as the systems work, and even higher amounts. There are two important implications. First, credit facilities and interest subsidy would be much better than capital subsidies. These not only lead to beneficiaries undermining the value of systems but also undermine the concept of service by the technology providers who move on after installation. Second, and it follows from the first, independent entrepreneurs would be dependent for payment on the functionality of systems.

They would, therefore, focus on service. Besides they would also be in a position to design products, even within a mini-grid, as needed by the different beneficiaries, not one size fits all.

Therefore, the subsidies proposed are best converted into interest subsidies. This would make the task of disbursal of subsidies also easier and transparent. The loan should be for a longer period of 12 years with a provision for one instalment of loan to be given after 4–5 years for battery replacement. There would be milestones to be achieved. When loans are given in this manner it would allow for proper audit of costs and returns and a decision can be taken later for future. Simultaneously, standards could be set, monitoring framework could be strengthened and accountability for proper quality of service should be ensured. The choice of the process should be left to the entrepreneur. And Banks should be encouraged to lend for such projects and a hassle-free system should be put in place. In fact they should have lending targets. What better way to improve the rural credit deposit ratio? This would also help drive the marketisation process.

The International Finance Corporation (IFC) estimates that the poor, energy-deprived people of the world spend nearly US$37 billion every year on low-quality energy solutions to meet their energy needs and that this represents a substantial and largely untapped market that the private sector—with appropriate government support could tap at a profit (IFC, 2012). Another study sees a US$2 billion opportunity for investment in India's energy access needs.[12] This could be a good example of the concept enunciated by the late Prof. C.K. Prahlad of enterprises benefitting from the bottom of the pyramid. Bihar was absent in RVE, and is largely absent in bank-financed systems. It is instead open to the private grey market with poor quality of products and poor service. It is reported that in its capital city Patna, a 1 km stretch on the Exhibition road is amongst the biggest off-grid solar market in the world, with business estimated to be over ₹500 crore annually. The real driver of this market has been the purchasing power of the people who want the benefits of electricity and go out of their way to try and get them and are prepared to pay for these, sometimes even relatively higher amounts of their incomes. There have been no subsidies. It is a testimony to the impact of solar, and its need, but also of our failure to make it into a genuine revolution through a proper programme, because this market could and should have been facilitated by a banking system and soft loans. This sadly reflects the story of missed opportunities.

Bilateral/Multilateral Agencies

This brings us to the role of bilateral and multilateral agencies. The Norway government fully funded several village projects with assistance from the MNRE, but because of that reason, they could not be business models. The UK government continues to assist projects but it is not clear how and where. GIZ also does certain projects as pilots, some on their own, which have not been successful. In September 2013, under Indo-US Energy Dialogue PEACE (Promoting Energy Access through Clean Energy) was launched. A fund has also been set up under this initiative to support the development of early-stage, innovative clean energy solutions for un-served and underserved populations, with equal contribution from India and the USA of around ₹25 crore for providing grant funding to develop and test innovative products, systems and business models.[13] We have to see how this will operate. A concerted effort is lacking and no models have emerged out of these efforts, and sadly, no entrepreneurs too.

The World Bank had effective solar programmes earlier in Indonesia, Sri Lanka and China but it did not approve the Million Solar Homes Initiative in 2002. It is currently preparing for a large project in some districts of Uttar Pradesh and Bihar (World Bank, 2010). It appears keen to attract large players to this area and is probably contemplating much larger systems which could be with one grid or with several mini-grids serving several thousand households. This may not be a practical approach as we discussed earlier. Currently they are getting Feasibility Reports prepared by different agencies. They need to do this quickly. It is, however, important to note that the very fact that they are proposing a VGF means that the subsidy structure of the Ministry does not make a project viable. The Asian Development Bank (ADB) has been talking about funding off-grid projects for the last several years but there is little actual progress. They are perhaps simply not geared for this work. Therefore, bilateral and multilateral agencies are simply not providing enough capital to make any real impact.

It is evident, therefore, that electricity access through renewable energy systems also face enormous challenges and many barriers. What is most important, however, is that the technology has become mature and cost competitive and there are really no fuel issues. This creates the opportunities which did not exist only till some time back for a large scale-up. Many of the barriers arise from lack of a conducive policy framework or lack of experience or building up of confidence and capacity. These problems can be overcome by organisational learning, capacity and confidence building, flexible and facilitating financial arrangements and availability of more funding and by slowly but surely creating an ecosystem which will develop into a market. Let us then look at the way forward.

An Alternative Framework to Achieve Universal Electricity Access

We have seen that the challenge to provide electricity access is immense. Solutions require recognition and understanding of the problems and a determined effort to address them. Above all it would require a fundamental reform of energy pricing. There is almost a unanimous view that rural pumping and household tariffs have to be raised to certain minimum levels. Urban tariffs have also to be revised, including much higher incremental charges for those who consume well above the average. There needs to be a gradual subsidy reform of kerosene. At the same time transmission (and theft) losses have to be brought down to less than 10 per cent in the next 3 years with suitable milestones. It is imperative that a national consensus be built on these issues quickly.

We can, of course, continue with the current approach to rural electrification, which by default seems to be the case. We can continue with more supply going to wherever possible without any specific targeting. In this case the reliability of supply, problem of evening supply, and continued theft and transmission loss will not be addressed. The position of the utilities will continue to get worse in proportion to the additional supplies they make. They will seek more subsidies, and perhaps bigger bail outs from the government. This is no longer sustainable. And perhaps we may be able to cover 50–100 million more people till 2030.

Challenges for Universal Electricity Access and Way Forward 145

Therefore, an alternative, or additional, approach based on renewable energy systems is necessary. The evidence in the form of case studies as detailed earlier suggests that renewable energy provides a viable and desirable economic option for rural electrification including the meeting of unmet demand in electrified villages.[14] However, the success stories are location specific and do not lead to a model that could be adopted universally. Also detailed figures about costs and returns have not been generally shared. The overarching requirement is of developing a sustainable energy access market that is over a course of time not dependent upon government assistance and that leads to increased penetration in rural areas with improved quality of supply and lower cost of delivery. But the essential precondition for developing this is first an acceptance, not grudgingly or as fait accompli only, that this is a viable doable option and perhaps the only way to achieve universal electricity access. All stakeholders, including state governments, have to first accept that centralised grid cannot achieve this, even substantially, and that renewable energy decentralised solutions are part of the mainstream solution, not a transitory exercise. Only then can the necessary policy changes be introduced, the required regulatory and institutional framework developed and a strategic action plan prepared and implemented supported by the necessary funds and banking support. Let us see what needs to be done and the way forward.

The choice of electricity service model for a particular village/location would depend upon the nature of demand and the nature of supply. There could be three broad choices.

The first approach would be augmentation of grid power availability to such rural areas, particularly with high grid penetration and low reliability of supply using locally available renewable energy resources feeding into the distribution grid or meeting captive loads such as the project in Vadodra. This would strengthen distribution and supply in existing grid areas where demand is high (say in the range of 100 kW and above up to 2 MW) and location is suitable. Power would be fed into the distribution sun stations which directly service consumption points. But it is necessary that the supply is not so unreliable that renewable energy generation is wasted because of grid dropping. This is expected largely to address industrial and commercial needs in rural areas. It can also meet rural agricultural needs where there could be clusters of pumps.

Locations could be rural peripheries of cities and towns, block headquarters and rural markets and along connecting roads. In the case of captive supply, there should be automatic permission and the parties concerned would fix the tariff. But feeding into the grid would require a tariff setting mechanism for the electricity supplied to the grid. Models of such projects exist. Hundred megawatt capacity of projects of size 1–2 MW were part of the first phase of the National Solar Mission. Where the load is largely day time solar projects could be tried, where longer we could go for biomass projects. This is only augmentation, rest power must come from the grid. These would be cases where VGF to reduce capital cost, preferably as interest subsidy, could be the right instrument which could reduce levelised tariff to ₹5 with some escalation factor. Otherwise the utility may not be interested in purchasing the power. The possibilities have been explored in the models—a network of such renewable energy projects feeding into the distribution grid can be set up across the country. In fact, that would be a very desirable option because of the many advantages of feeding power at the point of consumption. It takes care of transmission requirements and avoids losses and provides better quality power. Such projects can only be delivered by renewable energy sources. These projects would even be better than large solar projects in distant places. In such situations, the focus could be on increasing transformer capacity, instituting proper metering and billing and ensuring collection. Since electricity supply is likely to be more regular some tariff can be raised. There have been some pilots for improving all these issues in Singhram and Chord villages of Haryana which started in end 2012, which have improved both power supply and finances and simultaneously reduced both consumption and theft![15] REC and the proposed Rural Energy Authority jointly with the state governments can identify suitable locations. The authority would help implement them. There is a strong case for developing 200–300 MW in the next 5 years leading to a possible coverage of 2,000–3,000 MW by 2030. This may impact about 50 million people. Such an approach would require the early development of a policy for dedicated energy plantations. The advantage of biomass projects would be stable supply of power.

The second approach is the setting up of independent self-sufficient generation mini distribution network—micro-grids with renewable energy-based local generation to provide a village or a cluster of villages with electricity. Ideally, a mini-grid should service a village, but the

Challenges for Universal Electricity Access and Way Forward 147

entrepreneur should operate in a cluster of villages. In such situations, in most of the cases, only the basic and rural livelihood requirements are met, though there are some opportunities for commercial and micro-industrial usage.

The third approach is household electrification solutions—feasible to meet the basic level subsistence lighting energy needs of remote isolated villages where there is no aggregate demand or of electrified villages where supply is uncertain, especially in evening hours.

Solar home lights in India have been primarily led by the government although there have been many private efforts in the last few years. It has been suggested that procurement procedures and the remoteness of the work areas, which are far off villages generally, have not led to dissemination of high quality products nor to provision of effective after-sales services. The high levels of subsidy have not created ownership and militated against sustainability. Provision of solar lights was also seen largely as a dole. The requirements of subsequent battery replacement have also been really left unaddressed. Over the years, therefore, it has become clearer that solar programmes must be operated as a business, to take care of capital costs and they must generate sufficient revenues. These should enable the developer to service his debt, provide for proper service support, cover defaults, save for future requirements and get a decent return. So, if solar has to become a product, and not a commodity as entitlement, its commercialisation should best be left to private enterprise that should pursue this in their self-interest. The key policy objective must be to attract more entrepreneurs to the market and assist their growth so that they can provide this service and the numbers of customers grows. Subsidy, therefore, as a key instrument, needs to be designed to convince these entrepreneurs that there is an attractive opportunity to get a decent return and that they should remain in this area and further promote it. It is this consideration that should then decide the nature and levels of the subsidy. This subsidy should be available to both the developer who runs a mini-grid and the one who sets up a system to supply home lights. There will be difference in the amounts. We need to look at the earlier World Bank projects where grants were given to such systems operators who provided home light systems to consumers to enable them to set up their own distribution infrastructure. They could also be flexible in what they charged their different customers, lower for smaller systems and more for larger ones. This flexibility must also be permitted.

In the latter case, there is the additional need for consumer finance for the beneficiary which helps overcome the fundamental constraint of initial high cost of the technology into an operating cost thereby enabling modest and affordable instalments for paying over time. The corollary is that there should not be a separate parallel government-subsidised tender-driven market.

Selection of the best suited option amongst these would be decided by prevailing conditions and the institutional capacity. The kind of electricity service which could be the best possible option at a particular location has to be decided taking into consideration the area, habitat, availability of renewable resource, likely productive uses (present and future), status of power supply and affordability, etc. It would also be important to assess the likelihood of grid supply improving in that area. Hamlets at some distance from electrified villages may require special attention. Such an assessment will also give a preliminary idea about the WTP or alternatively what the villagers are spending at present on energy services. While undertaking such an exercise, the marginal cost of extending grid (if grid is not there or for augmenting it) would also need to be considered in conjunction with the overall economic benefits that accrue to the society and the economic cost of such an augmentation. The past experience and limited replication of renewable energy-based electricity access suggests that independent small-scale initiatives alone cannot lead to desired success levels. Public intervention would be necessary for organising the market and promoting sustainable technologies. A suitable framework with government support has to be put in place which will facilitate an entrepreneur-driven business model approach with innovative institutional, regulatory, financing and delivery mechanisms.

Actions Required

Institutional Structure

The first step would be to look at institutional arrangements. One of the major problems is that there is no separate institutional structure to look after decentralised renewable energy services. In the states the

Challenges for Universal Electricity Access and Way Forward 149

concerned entities are the state utilities, which we have seen are neither equipped nor interested in this. The existing national body is the Rural Electrification Corporation. This funds the infrastructure development for providing grid electricity to rural areas but has nothing to do with actual provision of services, except that, as a peculiarity, it has been asked to implement the DDG scheme. This was because this scheme has inexplicably been connected to the utilities. Rural electrification has been the primary function of the Ministry of Power and it is likely to stay that way, even though there will be substantial contribution from renewable energy sources in the future. Therefore, we need a completely new Institution like a Rural Energy Services Authority which will spear head not only renewable energy projects for electricity access but also improved stoves for cooking energy access and implement the family size biogas plant programme. This would have to be housed in the MNRE. Currently, energy access issues and programmes do not get the necessary attention, focus or budget. A separate authority will provide all three. Therefore, it is proposed that a commercial entity in the form of 'not for profit' Section 8 Company with government having minority stake (at least as an objective) that works in public–private partnership mode should be formed.[16]

The company would only work towards decentralised renewable energy projects and cooking energy. The main functions of the company could be (a) set up decentralised renewable energy projects or facilitate them; (b) fix minimum standards for such projects; (c) increase R&D efforts for standardisation of products, systems and services; (d) create an ecosystem for universal electricity access through renewables; (e) develop a cadre of entrepreneurs, including those at village level, for taking up electricity access projects in business mode; (f) determine the level of funding required for the projects make them viable; (g) cater to the human resource requirement through creation of an ecosystem for continuous capacity building; and (h) distribute subsidy to such projects. On the cooking energy side it would perform all the functions which have been described in Chapter 7 to help develop proper and long durable stoves; get studies and research done for such stoves and impact studies; do large-scale awareness; prepare and implement a big dissemination programme; disburse subsidy for the stoves; do monitoring and evaluation; prepare proposals for carbon markets and implement them.

This Authority would have an Advisory Board which would include representatives of Ministry of Power, Ministry of Health and Ministry of Rural Development. It will develop a strong collaborative relationship with states and help in strengthening this part of state renewable agencies. It could coordinate with existing DRDAs in districts. It would have a strong monitoring and evaluation wing. This Authority will be able to receive money from CSR, Bilateral agencies, Foundations, Government Budget, National Clean Energy Fund, MP LAD, etc. A decision would need to be taken whether it would also lend or provide interest subsidy directly or that function be assigned to rural banks with a clear cut direction.

There are such organisations in many African countries but they either do not have the autonomy or power or funds or direction to do the needful. They have, therefore, not been generally successful. Therefore, care has to be taken to ensure that this Authority has all these. The government has created a Solar Energy Corporation of India to facilitate implementation of grid and off-grid solar rooftop projects. It also distributes subsidy. We are convinced that the creation of such a Rural Energy Authority will provide the necessary focus and attention, and consequently impetus, to enable universalisation of energy access and must become a necessary first step, symbolic also of the government's new commitment to achieve this through renewable energy.

Comprehensive Mapping

The next step would be a comprehensive mapping exercise. The newly set up Rural Energy Service Authority, in coordination with the designated state nodal agencies for electricity access purposes, should develop each state's electricity access map wherein areas may be categorised as high grid electrified with satisfactory power supply and high household electrification levels; grid electrified with inadequate power and low household electrification levels; no-grid but planned in near future; and remote isolated villages or hamlets of electrified villages where there is no aggregated demand. In effect, except rural areas having satisfactory power supply, or likely to have this in the near term, there exists potential for implementation of the electricity access projects in all other areas. This kind of mapping will take care of areas where grid supply is or could be good, including the evenings which will avoid the unstated controversy of what happens if the grid comes. We had suggested a new definition of an electrified village. There can be variations regarding the number of households connected, the number of hours of power supply during the 24 hours and during peak evening period. This can give us mapping of areas with different conditions which could help us decide which of three options should be followed. But meanwhile we should also look afresh at areas covered by the RVE programme and the bank-financed programmes.

RVE Areas

If we take the statistical figure of villages covered under RGGVY, and add to that the villages covered under RVE, most of the villages would have been covered by now, with perhaps some hamlets, and some very small scattered houses, left out. But RVE started many years ago and it is time to think of what can be done in these villages if they have still not been electrified.

Since most of the systems were provided only in the last few years, it could reasonably be expected that the solar panels would largely be available, though some may not be in use for various reasons. However, there could be large-scale shortage of batteries, unless individual households have replaced them.[17] Therefore, it is now necessary to have a house to house census done of the beneficiaries and draw up a bank-assisted programme of rehabilitation based on possible categories of the state in which systems are found. This is possible because wherever systems are not working the beneficiary would be using kerosene again. This exercise would involve obligations of companies who were to maintain systems over a period of time. It would, however, be imperative that proper maintenance arrangements are made this time around with a cluster approach and a plan prepared for subsequent replacement of batteries. Ideally this exercise should include the study and development of a system and market for collection of old batteries and their replacements as well as disposal and the same for CFLs. This aspect perhaps has not been examined before and actually presents a separate challenge by itself, and one which needs to be comprehensively addressed. Wherever CFLs have to be changed, equivalent strength of LED lights may be offered. The time has also come for this shift to happen. People have to be told that the solar panel has to be maintained properly and that 15–20 years later this would need to be replaced also on their own, with suitable bank assistance. It is this requirement of continuous change of batteries, and of absence of guidance to an ordinary villager, that makes the alternative of privately managed mini-grids much more attractive. In areas where work was done many years ago, the batteries have not been replaced and some panels exist, a very radical approach would be that where the villagers desire to now have a mini-grid, the existing panels can be bought by the operator at discounted prices and used in setting up the grid system. This could be tried out first at a few places. Alternatively, we simply start with a mini-grid afresh. Naturally, the focus should be the villages covered earlier.

Bank-financed Programmes

What should be done in this context with the bank-assisted programmes where the target under the National Solar Mission is to cover 20 million households by 2022, but where progress appears to be relatively slow? The main problem remains for the many electrified villages where a large number of households have not taken connections. Here the mini-grid may be the preferred option. The bank-financed subsidy programme for solar home lights would essentially cater to such of these areas where the position is relatively better, and more households have taken connections. Because of the bank–solar company partnership, it is expected that this would be more sustainable too. It can also be reasonably expected that the battery change requirements would be adequately met with perhaps a small unsubsidised loan later on. Nevertheless, there is a strong case for doing a 10–20 per cent check to see what is happening on the ground, particularly for villages where financing was done over 3–6 years ago to see what happened and is happening and what is possible in some areas where banks have not been very active.

The question to ask is that if bank financing is such a good model then why the progress is so slow when the need is so obvious and the business case established. The primary reason, as discussed earlier, appears to be that the banking community in general is simply not coming forward for various reasons. The first issue relates to inertia—why go into a new area which would add to the workload, and could be risky, when no one is really prodding you. This is part of the risk-averse governance ethos of our country where routine is king, and initiative and innovation are neither recognised nor appreciated. The Finance Ministry is also not pushing. All the pressure comes only from the MNRE. Second, there is a worry about the credit worthiness of the borrower. Will people return the money? Third, there are high transaction costs. The number of beneficiaries will be very large but the individual amounts would be small. Perhaps extra personnel should be allowed to these banks based on certain norms of lending and this cost should be reimbursed to them. It is surprising that the Rural Bank-financed programmes are limited to just a few banks and areas—parts of Uttar Pradesh being predominant and a couple of other states, which surprisingly includes Andhra Pradesh where both electrification and household connectivity are good (Figure 6.1).

What is different with these banks and areas? This is a question for which so far there is no answer, and perhaps that has not even been asked. The leaders of those banks who started this initiative deserve full credit.[18] However, this does not appear to be seen as a priority area either by NABARD or the Department of Financial Services. If we are in mission mode, and this lending also becomes a target of

Figure 6.1
MNRE subsidy for bank-financed projects—No. of units

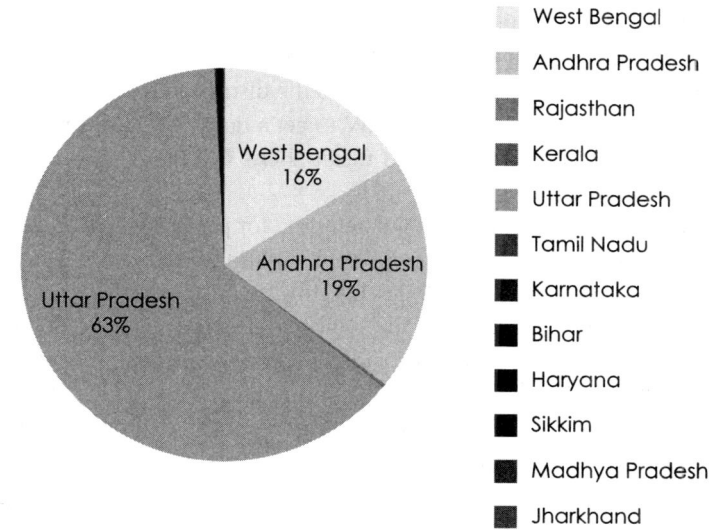

Source: Ministry of New and Renewable Energy.

that Ministry, then things may change. The experience of repayment so far has been excellent, though CSE field visits mentioned earlier talk about defaults in some places becoming worrying—some linked with service issues. We need more clarity on this account and a study should immediately be carried out. A system that reduces transaction costs for the beneficiaries as well as the bank branches (and allowing for repayment also on day basis) would be desirable. The government would need to work closely with the banks for an intensive awareness drive in the concerned villages. The MNRE had already started a scheme to fund banks for this purpose. If this approach is followed, which is necessary, several million households can be covered in a reasonably short while. One of the major advantages of this approach is also that it does not come into conflict with the grid. The only requirement should be that consumers with electric connections should be billed on their actual consumption and/or not have to pay high fixed charges every month to the utility. Bihar has always complained about the adverse rural credit deposit ratio. What better way than this to redress the imbalance somewhat?

Requirement of Funds

Let us, therefore, separately think of meeting the lighting needs of around 50 million rural households currently dependent upon kerosene.[19]

In the first case of projects feeding into the distribution grid we could presume a VGF of ₹2–3 crore per MW to get a tariff of ₹5 per unit. For 3,000 MW in 15 years this would mean about ₹7,000 crore or about ₹500 crore annually.[20]

Suppose an amount of ₹12,000 is required for providing a household with a reasonably good quality solar home lighting system with 2 lighting points, a small fan and mobile charging facilities, whether through mini-grid or solar home lights, the total investment would be 60,000 crores. If 40 per cent subsidy is provided by the government, this would require ₹24,000 crore. The rest would come as bank loan and beneficiary contribution. Each household would be very easily able to pay back the loan based on their current kerosene consumption cost, and paradoxically enough, this would become more viable, if the public distribution system (PDS) kerosene cost is revised. And after a few years they should also be able to replace the battery by themselves.

If this process is theoretically completed over 15 years, it would require only ₹1,600 crore per annum to cover 3.5 million households every year. At ₹34 current subsidy per litre, for kerosene, and saving of 25 litres annually per household, for 3.5 million households this would imply savings of ₹300 crore in the first year and additionally for each year. Over 15 years, the subsidy of over ₹40,000 crore alone would be saved other than incidental benefits. Should we not encourage rural investments and should we not discourage use of kerosene? Report of the High Powered Committee on Financial Position of Oil Companies had also made an important recommendation that all those rural households below the poverty line (about 50 million) be provided with one solar lantern costing US$75 each.[21] Since now better solar alternatives are available, we should provide even the poorest with more light. Mini-grid or solar home lights are the answer. Costs would be more than manageable. We could afford to be liberal with solar subsidies for rural areas.

Success of electricity access programmes would depend upon how effectively the issues relating to skewed market, financial/economic inadequacies, and infrastructural bottlenecks are addressed. The Government of India's role would be to act as enabler and facilitator and also institute a mechanism to link various stakeholders. Further, energy-enterprises-driven electricity access programme would be based on pull approach wherein demand for action emanates from the entrepreneurs or other stakeholders. The role and responsibility of the government would be of ensuring smooth flows of required inputs to the enterprises.

Energy Access Trust Fund

We have made a rough estimate of funding requirements. In order to undertake universal electricity access in a speedy and targeted manner, there will be a need for pooling resources from all possible sources. The best way for achieving it would be to create a Trust Fund which should be able to accept contributions from Individuals, Companies, Foundations, National as well as International Agencies, which are philanthropically oriented and have dedicated funds for supporting such interventions, and also from governments (central/state), plan grants, diverted fossil fuel subsidies, especially of kerosene, budgetary allocations, and diverted energy subsidies. In addition, there are bilateral and multilateral mechanisms, which support such interventions.

There is another way to raise resources. Assuming a normative figure of 40 per cent electricity being consumed in the urban areas by the highest incremental domestic consumers and commercial establishments a surcharge restricted to even 2 paise per unit of electricity could collect an amount of ₹600 crore per year. The obvious and primary source of funds would be the National Clean Energy Fund, unless it is diverted for other uses. There possibly cannot be a better use of this Fund than Energy Access. The new government has also doubled the cess on coal so that now annual accrual would be more than ₹7,000 crore. Creation of such a Trust Fund will provide necessary functional freedom for providing assistance to the electricity access projects and will contribute to the payment of operational incentives to the entrepreneurs and ensure sustainable operations of the distributed energy systems. It will be located in and administered by the proposed Rural Energy Authority.

As already mentioned, there was a proposal for an Energy Access Fund where the governments of Norway and UK were willing to put in substantial amount of sums with an equivalent contribution from Government of India. But, it did not get approval. That fund would have met the VGF needs or provided loans at low cost, even lower than 5 per cent per annum. But it was not to be. That proposal could be resurrected. The present president of the World Bank, Jim Yong Kim, is restructuring the bank and preparing it for achieving two targets—abolition of extreme poverty by 2013 and boosting prosperity for the bottom 40 per cent of the people in developing countries irrespective of income. Hopefully, energy access projects will get more attention and funds. Hopefully also it will now develop a market place for voluntary carbon credits for improved cookstoves. World Bank lending can collaborate with such a fund, while the carbon revenues can become part of the operations of the Fund. In fact, it has been suggested that the bank house and manage a fund to target 100 million solar homes by 2025 (Miller, 2009). This may be ambitious but a target of 50 million should be planned. This will happen if it is taken up as a global mission. It has the earlier successful examples to follow. It should follow that route rather than the bidding route it is proposing which may not succeed. This would also certainly be amongst the best use of the proposed Climate Fund, which is still to see light of day. In fact, if we hive off a small window for energy access from that Fund it may even get operationalised, and that too fairly soon.

Challenges for Universal Electricity Access and Way Forward

Corporate Social Responsibility (see Box 6.1)

It would also be appropriate to discuss here what can be done through grants of non-profit organisations and through CSR funds. So far there are limited pilots. CSR funding tends to be *ad hoc*, not valuing long-term involvement and building sustainability. Thus, they may distribute lanterns or street lights or even some home lighting systems, as Container Corporation of India (CONCOR) has done for about 1,000 households near Jodhpur through Central Electronics Limited (CEL). It would be better to help set up mini-grids or solar charging stations in a cluster over time. These one-time supplies are normally not very sustainable as we are seeing in the RVE programme. We need to change this and use these funds strategically. CSR funding could have an important catalytic role to play in areas where markets have yet to be created and government is finding it difficult. One of the problems is that the public sector does not go to the private sector for implementation. Rather than individually

---

Box 6.1 Corporate social responsibility

Section 135 of the New Companies Act 2013 mandates that every registered company shall ensure that company spends at least 2 per cent of its average profits during previous three financial years as a part of Corporate Social Responsibility (CSR). The identified activities include eradication of hunger and poverty; promotion of gender equality and women empowerment; and ensuring environmental sustainability. Renewable energy development and deployment is inclusive in all these areas. Broad estimates indicate that more than 2,500 companies will be covered under Bill's CSR mandate and there could be an annual CSR budget of around ₹15,000 crore. Further, the Department of Public Enterprise issued consolidated guidelines on Corporate Social Responsibility and Sustainability for Central Public Sector Enterprises (CPSEs) that came into effect from 1 April 2013. These guidelines provide a framework for CSR activities by CPSEs, in a broad spectrum of area including for leveraging green technologies, and also for contributing to inclusive growth and equitable development through empowerment of the marginalised and underprivileged sections/communities. In 2010–11, profit of profit-making CPSEs stood at around ₹1.25 lakh crore—this makes around ₹2,500 crore for CSR activities.

Source: Ministry of Corporate Affairs, The Companies Act, 2013; PHD Chamber and Ernest & Young: Corporate Social Responsibility in India: Potential to contribute towards inclusive social development; and Department of Public Enterprises, Ministry of Heavy Industries and Public Enterprises.

funding projects they could partner with the proposed authority and make strategic donations to the proposed fund, and the authority can get the project implemented in the chosen place. In fact, we can be more liberal with the grant portion as subsidy than normal, as otherwise, CSR would have resulted in full grant which is not desirable. Electricity access projects are socially extremely relevant, with high positive social and economic externalities and meeting the overall human development needs. They are also short of funds. Therefore, this is really one of the best areas for CSR. Minda is using such funds from public and private companies. There is a need to dramatically up-scale.

The companies which should be specially interested are all who are involved in the power, mining and oil sectors. These systems are going to replace kerosene and diesel. The most needed areas are those which are near the mines, which generally have been the most neglected, apart from Bihar and eastern Uttar Pradesh. So such companies have a special responsibility. And many of these are in the public sector.[22] Their CSR funds can alone help to cover thousands of households every year. Even ₹500 crore annually could make a big impact. Other companies which should be interested are those which see rural areas as markets for their products like TVs or lights or batteries. They can help develop models, entrepreneurs, studies, etc., which will be very useful as a full roll out happens. What would be ideal would be the willingness of MPs and MLAs to use a part of their annual constituency development fund along with, or separately, to promote sustainable energy access projects in some villages every year in their constituencies—even 5 per year would do. One MP, Mrs Rebello tried to do this in Jharkhand. In fact, she wanted the entire Albert Ekka Block in Gumla district to be electrified in this manner—with CSR from SAIL, her own MP fund and subsidies from the government. A proposal was also sent to the Clean Energy Fund, but it was not approved. It was of course ambitious, but it shows what can be done at more realistic levels.

Entrepreneurship Development

As we have seen earlier, developing a pool of entrepreneurs and creation of entrepreneurship-based business models that encourage productive use of electricity at the household level will be fundamental to electricity access. Entrepreneurs, from the local communities, would be expected

Challenges for Universal Electricity Access and Way Forward 159

to establish energy micro-enterprises to produce and to distribute energy carriers to rural households at an affordable cost. The strategy to create local enterprises would rest on skill enhancement; incubation; and facilitation of market linkages. Entrepreneurs will have to be provided with performance-based incentives and penalties that are similar to grid-connected renewable energy projects, but cognisant of the operational difficulties in rural areas and environment. The incentive levels should be adequate for viability of a project.

Absence of entrepreneurial and technical skills would be a major hindrance to setting up of energy enterprises and therefore prospective and existing entrepreneurs will need to be trained on various aspects of techno-economic and commercial viability of enterprises: technology, management, finance and government policies. Essentially they must learn to run a solar business properly. The involvement of relevant technical institutions is essential for training these entrepreneurs. The existing training infrastructure created for training of rural franchisees under RGGVY programme could be enlarged in scope and renewable energy–based electricity access could be covered to enable the existing and prospective entrepreneurs to establish and run the rural energy enterprises. Appropriate training modules would need to be prepared. Already organisations such as Shakti Foundation have started such efforts. This exercise needs to become broad based and spread out to cover hundreds of potential entrepreneurs. This training in any case will not go waste. There are already programmes of the government where different kinds of trainings are imparted. There is a case for coordination. A mapping of possible entrepreneurs should also be carried out in high potential areas. A sound business plan would be a prerequisite for any entrepreneur to start electricity access business. A template of prototype business plans emphasising the profitable nature of the projects would serve as reference document for a prospective entrepreneur.

ORF in its report has very pertinently suggested that

> Companies such as SELCO and Husk Power who have been training entrepreneurs for the last several years could be invited to design a programme to train the trainers, who in turn could be act as trainers across the country to kick start a massive capacity building effort. It is very important for this program to reach the interiors of the country. There are over 10,000 ITIs, so the potential of this activity is huge. It would be challenging to recruit

trainers and thereafter potential entrepreneurs on such a large scale. This exercise can be ramped up over several years, as Solar PV applications continue to percolate into rural and remote areas and the demand subsequently increases. (Deorah and Chandran-Wadia, 2013).

Renewable energy systems have the potential of developing the rural entrepreneurship spirit which may lead to development of other agro industrial enterprises so that there is greater agricultural produce, more local processing and less wastage. In this context, it should be noted that renewable energy systems can also power cold storages and small dedicated food-processing enterprises.

Preparation of Feasibility Studies of Clusters of Villages

As the mapping exercise is done as proposed, there could be preparation of feasibility studies of clusters of villages which in fact could then become readily available. They could be used as training exercise as well as almost ready-made DPRs. The World Bank has started its own process but this exercise could be easily funded by the government and possibly the bilateral agencies could also support this. This could be a very important preparatory activity and will also help prioritise areas.

The proposed framework contains recommendations for the adoption of an integrated energy access programme that has inputs from the existing and ongoing energy programmes and includes new approaches and mechanisms to enable achievement of universal energy access in the shortest possible time frame. It, in a way, will complement the proposed global target of universal energy assess by 2030.[23] Rural Energy Service Authority along with the Trust Fund is expected to bring about transformational changes in both *process* and *progress* of designing and implementation of energy access projects. Further entrepreneurship-based models would ensure sustainability in long run and also create large number of rural jobs in a decentralised manner. The approaches suggested would be win–win situation for all.

Notes

1. In a visit of Deepak Gupta to such systems in Bihar in 2012, women's groups expressed prima facie inability to manage such systems, though solar could be done, especially if it is a solar lantern charging station.

Challenges for Universal Electricity Access and Way Forward 161

2. Discussion of Deepak Gupta with CMD IREDA in April 2014.
3. Rural Electrification Corporation. Available at http://rggvy.gov.in/rggvy/rggvyportal/plgsheet1.jsp. Accessed on 23 November 2014.
4. Policy and Regulatory Interventions to Support Community Level Off-Grid Projects: A Study for CERC. Available at http://www.shaktifoundation.in/cms/uploadedImages/final%20report_cwf_off%20grid_nov%202011.pdf. Accessed on 23 November 2014.
5. Ministry of New and Renewable Energy (2012): draft No. 13/14/2011-12/RVE, Remote Village Electrification Programme.
6. MNRE order dated 30 May 2013. Available at http://mnre.gov.in/file-manager/UserFiles/amendmends-benchmarkcost-aa-jnnsm-2013-14.pdf. Accessed on 23 November 2014.
7. MNRE draft no. 13/14/2011-12/RVE, Remote Village Electrification Programme (March 2012). Available at http://mnre.gov.in/file-manager/UserFiles/village_lighting_programme_scheme_2012_13_for_comments_feedback.pdf. Accessed on 23 November 2014.
8. Based on the project costs of DDG projects and discussion with REC officials. The average cost would be ₹320/watt but project-wise detailed break-up is not available. It is understood that support level was not uniform across the projects. A list of REC projects has been given in Chapter 5.
9. There was a first time local entrepreneur in Buxar who wanted to start rice husk based gasifier projects in several villages in 2009. In spite of many meetings with SBI, including with Deepak Gupta, the then Secretary MNRE, who wanted to try this out as a grass root model, they refused to lend. They were quite apathetic in their attitude, and even somewhat patronising. Sadly, references to the SBI headquarters went without response. There were other issues like the change of land use which the local administration could not do.
10. It is also easy to emphasize the risks for such small projects even as bankers become much less stringent where large projects are concerned. Witness the huge NPAs of large companies, which alone would be more than the total credit requirements of decentralized energy systems! Indian banks have written off ₹1 lakh crore over a period of 13 years. Source: *Business standard* 19 November 2013. Available at http://www.business-standard.com/article/finance/forget-leaky-pds-it-s-the-banking-system-that-needs-a-fix-113111900635_1.html. Accessed on 22 November 2014.
11. RBI Circular No. RBI/2012-13/138 (RPCD.CO.Plan.BC 13/04.09.01/2012-13) dated 20 July 2012.
12. Power to the People: Investing in Clean Energy for the base of the Pyramid in India. Centre for Development Finance and World Resources Institute. September 2010. Available at http://pdf.wri.org/power_to_the_people_front.pdf. Accessed on 22 November 2014.
13. US Department of Energy. Available at http://www.energy.gov/articles/us-india-energy-dialogue-2014. Accessed on 22 November 2014.
14. The World Bank (2010) report 'Empowering Rural India: Expanding Electricity Access by Mobilizing Local Resources' specifically focuses on the issues and concludes the same.
15. More details are given in World Bank (2014).
16. An effort was started in 2010–11 in MNRE to set up an Energy Access Centre. But it did not go on to its logical conclusion.
17. MNRE's feedback gives mixed results.

18. Deepak Gupta, the then Secretary, MNRE had held many meetings with the banks in Bihar, Jharkhand and Odisha and other areas of Uttar Pradesh in 2010–11, but there were always limited and half-hearted responses, and consequently there was little progress.
19. Although the households dependent on kerosene for lighting are around 72 million, it is assumed that around 22 million will be covered in the normal course.
20. These are broad estimates. Detailed calculations would be required. As the tariff from other sources increases, the VGF would decline. Additional support would be required for developing dedicated energy plantations. But in case of biomass projects, tremendous benefits will accrue to the concerned areas from supplies of fuel.
21. Report of the High Powered Committee on Financial Position of Oil Companies (2008). Available at http://www.indiaenvironmentportal.org.in/files/B%20K%20Chaturvedi%20Report.pdf. Accessed on 22 November 2014.
22. Major Public Sector Companies include SAIL, Coal India, NMDC, NALCO, ONGC, NTPC, IOC, RINL, etc.
23. UN Secretary General's 'Sustainable Energy for All' initiative that seeks universal energy access by 2030.

7
Access to Cooking Energy

The Problem

Ever since man discovered fire and started eating cooked food, he has used three stones and a collection of branches, twigs, etc. to light the fire. Over time, the hearths became more sophisticated as housing developed and in the last century fossil fuels became available for clean and easy cooking. But should the early man descend to earth again, he would surely not recognise where he has come until he stops to share a meal with anyone of over 3 billion people who still rely on biomass (wood, agricultural residues and animal dung) for cooking. It is indeed a wonder of our times that with all the dramatic progress made, so many still cook their food under those conditions. But the greater part of the tragedy is that the world has not done much about it, and it seems that it also may not.

The poorest households in many developing countries still use three stone fires for cooking, although as you go up the income ladder, there are also traditional mud stoves or metal, cement and pottery or brick stoves, without chimneys or hoods. Biomass fuels in these households are mostly used in these inefficient, poorly vented combustion devices which result in the bulk of the fuel energy being emitted and wasted as products of incomplete combustion. The high moisture content of the biomass resources used and the low efficiency of the combustion process produce a lot of smoke and emit large amounts of particulate matter, which causes high levels of indoor air pollution in small and poorly ventilated kitchens. These are breathed by those living in the home, which is called exposure, leading to very detrimental effects to the health of family members, particularly women and young children, who spend a lot of their time in or near the kitchen. Burning solid fuels produce levels of air pollution that far exceed the health-based standards for safety in the household, especially with repeated episodes of intense emissions (IUATLD, 2009).

There are other problems as well. Solid fuels are time consuming to cook because it takes more time to get the fire going, more time to cook the meal because of low efficiency, and more time to clean afterwards on account of soot deposition. Collection of biomass takes substantial time, which could be usefully used otherwise and there is also drudgery and difficulty of collection, and sometimes issues of safety. There are also issues of loss of vegetation and forest cover, especially where populations are more and the cover not so much. It appears that the impact on health; the loss of time; the value of the environment, both inside and outside the house, have not bothered us much over time. Let us consider these issues in some more detail.

Health Impacts

Adverse Health Consequences

The largest single source of health impacts of energy is associated with indoor air pollution resulting from the use of biomass in inefficient stoves by poor people in developing countries. The most dangerous parts of the smoke are the very fine particles that are produced by burning of solid fuels that are able to pass deep into the lungs when a person breathes the air containing the smoke. Up to half of the total exposure in women who cook with solid fuels comes from the high levels of smoke produced when they are close to the fire, especially when starting or stirring the fire, and the duration of such exposure is long.

The number and the quality of available studies associating exposure to solid fuel combustion with respiratory diseases are limited but growing. Two types of lung disease have been judged to have strong evidence of association with such exposure—acute lower respiratory infection (ALRI) in young children and chronic obstructive pulmonary disease (COPD) in women. An analysis of the results of three studies suggests that this exposure also results in worsening of asthma with children having a 60 per cent greater risk and adults having a 20 per cent greater risk of asthma attack in the presence of biomass smoke. A study showed that elderly men and women living in households using biomass fuels are much more likely to have asthma compared with those using cleaner fuels. International Union Against Tuberculosis and Lung Disease (IUATLD) studies in India show that as many as 5 lakh premature deaths annually in children under five and women could be attributable to the indoor pollution caused because of the use of biomass (Smith, 1998). Another study suggests in addition 500 million cases of illness, which would cause huge morbidity problems and expenses in India's underserved rural health system (Von Shrinding et al., 2010). Emerging evidence suggests that indoor air pollution (IAP) increases the risk of other child and adult health problems, including low birth weight, perinatal mortality, middle ear infection, blindness through cataract and cardio vascular disease (Ekouevi and Tuntivate, 2013). Some of the effects have been given in Table 7.1.

Who Are Affected?

A significant proportion of those affected are infants and young children who have to spend a lot of time indoors and are being exposed to daily pollution in the form of small particulates that exceed WHO recommended limits by 10 to 50 times. A meta-analysis of all global studies indicates that the risk of acute respiratory infection (ARI) for children exposed to indoor smoke is 2.3 times higher than that for other children (Barnes et al., 2012). Young children are more susceptible than adults to absorb pollutants, since their lungs are not fully developed, until they reach their late teens. Cooks, almost wholly women, are also major sufferers. Old people are also affected since they spend quite some time inside the house (Figure 7.1).

Impacts

In its World Health Report 2002, WHO had estimated that indoor air pollution from household use of solid fuels was the fourth leading health risk in developing countries leading to high mortality. Another study argues 'disability adjusted life years' associated with indoor air pollution

Table 7.1
Health effects of biomass fuel use in cooking

Processes	Potential Health Hazards
Production	
Processing/preparing dung cakes	Faecal/oral/enteric infections, skin infections
Charcoal production	CO/smoke poisoning, burns/trauma, cataract
Collection	
Gathering/carrying fuel	Trauma, reduced infant/child care, bites from venomous reptiles/insects, allergic reactions, fungal infections, severe fatigue, muscular pain/back pain/arthritis
Combustion	
Effects of smoke	Conjunctivitis, blepharo conjunctivitis, upper respiratory irritation/inflammation, acute respiratory infection
Effects of toxic gases (CO)	Acute poisoning
Effects of chronic smoke inhalation	Chronic obstructive pulmonary disease, chronic bronchitis, cor pulmonale, adverse reproductive outcomes, cancer (lung)
Effects of heat	Burns, cataract
Ergonomic effects of crouching over stove	Arthritis
Effects of location of stove (on floor)	Burns in infants/toddlers/children

Source: Batliwala and Reddy (1996).

is a problem that concerns all developing countries to varying degrees (Lvovsky, 2001). Another study has estimated that IAP results in about 2 million premature deaths, or about 42 million person-years of life (DALYs) lost per year, due to respiratory and other diseases women and children.[1] The first ever estimates emanating from the largest study so far of the contribution of different risk factors to the Global Burden of Disease 1990–2010 found such annual deaths from breathing emissions due to burning of wood, brush, dung and other biomass for fuel leading to respiratory illnesses to be 3.5 million annually along with 0.5 million

Figure 7.1
Incidences of respiratory symptoms for males and females by age group

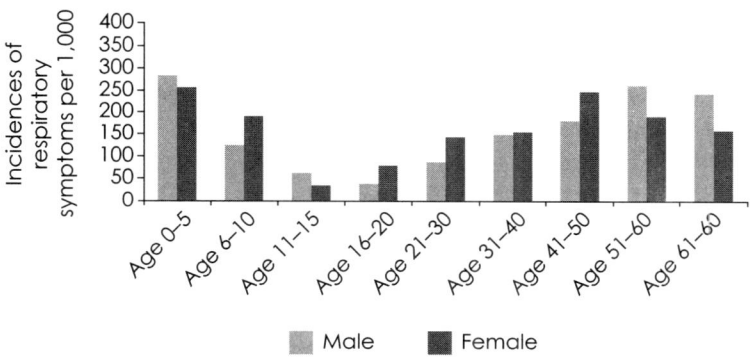

Source: Ekouevl and Tuntivate (2013).

from outdoor air pollution due to use of household fuels. This is significantly higher than what WHO had estimated.[2]

The findings from this study must draw our attention to several important facts. First, smoke from cooking fires was found to be the largest environmental threat to health in the world today. Second, with estimated pollution levels in Indian rural kitchens being more than 30 times higher than recommended levels, household air pollution from solid fuel use has been found to be the highest health hazard for India, with tobacco and blood pressure following. Third, this risk factor is far greater than poor water and sanitation, and some others. Finally, the number of premature deaths from such household pollution is greater than the number of premature deaths from malaria and tuberculosis (Figure 7.2) (IEA, 2010, p. 14).

However, this has not generated that kind of concern that those communicable diseases have. With better diagnosis and treatment and other preventive measures regarding other killer diseases, and little action in respect of rural cooking energy practices, the relative problem will only becoming worse. It is heartening to note that India's Health Ministry has recently set up a national target of 50 per cent reduction in exposure to IAP by 2025. We will have to wait and see what it does about this.

Figure 7.2
Premature annual deaths from household air pollution and other diseases

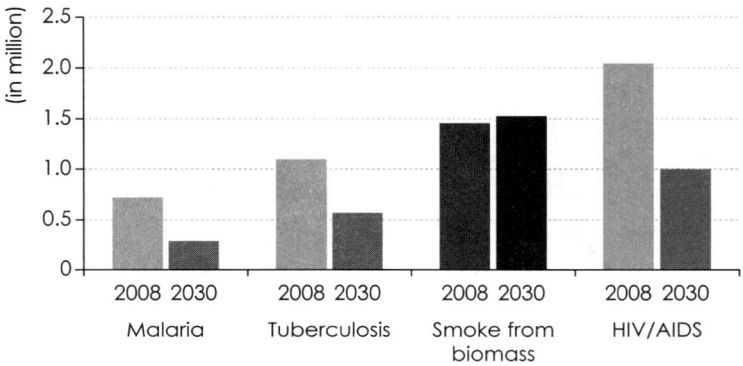

Source: Mathers and Loncar (2006) as quoted in IEA (2010).

Improved Cookstoves and Reductions in Pollution Levels

There are many studies that show worthwhile reductions of IAP by using cleaner fuels or improved cookstoves. However, they also show that well-functioning improved stoves show post-intervention levels well in excess of WHO Air Quality Guidelines (AQGs); that personal exposure is generally not reduced proportionately as much as ambient pollution for any given intervention; and that larger reductions would be more difficult to achieve with solid fuel interventions than with cleaner fuels. Nevertheless, even though intervention-based evidence of impact on health outcomes is still limited, there are indications that the degree of exposure reduction achieved with well-functioning improved stoves can still result in important reductions in incidence of child pneumonia and COPD (Bruce et al., 2011). Therefore, this intervention is not only well worth doing but also necessary.

Public Health Research

WHO has emphasised the importance of ongoing and planned intervention/research projects on the impacts of indoor air pollution on a range of health outcomes. For a sound policy on cookstoves, it will

> **Box 7.1 Cooking energy: Research agenda**
>
> Given that solid fuels are expected to be the main household cooking energy source for more than 3 billion people up to and beyond 2030, research is needed on improving combustion efficiency (to reduce emissions and save fuel), heat transfer, and ventilation of stoves. MNRE had proposed to the X prize in 2010 to have an international competition. But they backed out later. Perhaps India can have a national competition. We must find a way and fund improvements and number of organisations should be encouraged to do so.
>
> In view of limited understanding by the beneficiaries of better health impacts, health outcome research should extend to the intervention based evidence on health impacts required to strengthen advocacy efforts, together with further description of the exposure–response relationship for childhood pneumonia. In this context, we need further research into what different levels of IAP cause what different negative outcomes as also what changes can occur with different levels of reductions, with and without other factors which may influence outcomes.
>
> Economic evaluation is at an early stage, and there is a need for more empirical evidence on costs and the full range of benefits associated with various interventions across a range of settings. It is also important to determine the direct impact of household energy interventions on poverty reduction and the pathways by which this is mediated.
>
> *Source:* The authors.

be important to evaluate various health and broader impacts of interventions to reduce indoor air pollution to generate country specific evidence for making sound policy recommendations. Unfortunately, ICMR and other public health research institutions have yet not prioritised this research.

Other Effects

Effect on Forestation

There does not seem to be a real study regarding the impact of traditional cooking practices on deforestation but there is abundant evidence to suggest that extensive collection of biomass energy puts significant pressure on local resources. Anecdotal evidence also suggests that deforestation is indeed a threat to many under forested areas.[3] In general, the problems of collection, in terms of time taken and drudgery of collection, are only increasing. That is why the National Mission for Green India under National Action Plan on Climate Change had proposed expansion of services for cleaner cooking fuels, using of improved fuel-efficient biomass cookstoves and setting up biogas plants, etc., in areas bordering forests. It seems, however, that not much action has been taken in relation to that target. In the context of our discussion, this is not surprising. But it could perhaps be a useful exercise nationally to try and identify areas which may be facing the severest such threats. It is now widely acknowledged that the clearing of land for arable and pastoral agriculture is the main cause of deforestation, but there are some areas in sub-Saharan Africa that have shown devastating consequences. The focus of earlier cookstoves was on saving wood. Perhaps the evidence that generally this does not make such a great impact on deforestation may have actually diverted attention away from the need to have improved cookstoves. A study also calculates that if a programme of 150 million stoves is implemented, 196 metric tonnes (MT) of solid fuels would be saved annually of which 95 MT would be wood and 6 MT would be coal, both desirable objectives. The same study also estimates reduction by 4 per cent of India's total estimated greenhouse gas (GHG) emissions and 0.15 MT of annual reduction of Indian black carbon emissions which is approximately one-third of the total national human caused emissions of this pollutant (Venkataraman et al., 2010). Therefore, an improved cookstove programme would still be justified on this ground also.

Collection and Cooking Time

Issues relating to time taken for collection of biomass and cooking have also been studied. The evidence indicates that rural women in South Asia with access to improved cookstoves during the time when programmes were in full swing saved a significant amount of time in collection and cooking. They also consequently contributed to lessening of the amount of drudgery women suffer. One could even conclude that actually the value of all unvalued time spent in collecting fuel wood could easily pay for an improved stove that saves fuel, and thus collection time (Barnes et al., 2012). A study on linkages of energy with gender suggests that fuel wood collection takes its toll on human resources as household members typically walk 30 km and spend 41 hours collecting fuel wood a month. This is largely the responsibility of women, although older men are often also involved (Parikh, 2005). A study that reviewed economic benefits with various cooking alternatives suggests that in net present value terms improved stoves have the greatest overall benefit to society (Bruce et al., 2011).

In another study conducted by Deutsche Gesellschaft für Technische Zusammenarbeit (GTZ) for their Rocket stove programme in Uganda, and also consistent with a WHO study, benefits were dominated by fuel saving (including time collecting fuel) at 51.9 per cent, with time-saved cooking providing 14.3 per cent and health benefits 7.4 per cent. As such, improved stoves could provide much of the perceived benefits at very little cost (Bruce et al., 2011, p. 70). Time and fuel savings were more important, compared to health, even more so where averted mortality was not included among the benefits. The findings strengthen the case for interventions and for promoting awareness of the time and fuel savings in addition to health benefits. While women consumers may feel their time and energy is important it appears that it is the policy makers that have not given this any real value.

IAP and Poverty

It is evident that IAP from solid household fuel use is associated with a substantial global health burden for the rural poor, and in the case of India, this has been very clearly established. Addressing this problem, therefore, must normally be a public health priority. But we must also consider the organic relationship between reliance on traditional fuels and poverty, which, as in the case of electricity access, also operates in both directions. On the one hand, poor households do not have the financial resources and income security required to switch to more efficient and cleaner fuels and energy technologies. At the same time, the consequences of using traditional fuels—which include impacts on health, women's time and opportunities for income generation—add to the constraints on families attempting to escape from poverty. It is well established that health-related morbidities are by themselves also a critical causal factor for families slipping into indebtedness and poverty. For all these reasons, poverty is a key underlying issue in any consideration of rural household energy solutions for developing countries. Improving access to cleaner and more efficient household energy makes direct contributions to both health improvement and poverty reduction. Therefore, economic growth or other direct poverty reduction measures by themselves will not improve energy choices for poor households. Direct intervention for energy solutions will be necessary. This is similar to other needed household interventions such as clean water and sanitation, which serve to assist poor families' health and their prospects for improving their condition in other ways.

Policy Implications

Lack of Policy Response

This interlinked problem of cooking energy, and its linkages with poverty and gender and child issues, and its huge extent, would suggest that making access to culturally acceptable, clean and efficient cooking fuels should have been a priority policy concern as also a high public health priority. But somehow this seems to have been neglected over the years in almost all developing countries, as also in any worthwhile global action. These intertwined problems, unnoticed for centuries and first identified several decades ago, were put before the international development community as the 'other energy crisis' (Eckholm, 1975; Barnes et al., 2012, p. 1). Unfortunately, this crisis not only persists, but has perhaps only deepened, but even today it remains largely invisible to policy makers (Barnes et al., 2012, p. 1). The IUATLD has put this feeling in rather strong words: 'Such knowledge demands action—it is unethical to know about a problem and do nothing to attempt to address it. It is particularly scandalous when those most at risk are the most vulnerable members of the society, such as young children, the poor and the ill' (Barnes et al., 2012, p. 1). It is also a wonder that the health sector across the globe has not responded to this issue at all. This kind of non-recognition of the problem raises the question whether it is because it is a problem almost entirely of the poor and poorest?

Lack of Gender Response

The further surprising part is that there has also not been any real concern shown by civil society activists nor even as a gender issue of significance. There is so much discussion of gender, but the time and effort spent by women in using biomass for cooking has hardly any recognition. A group of female economists from India took note of this neglect some time back and called for a policy response. This is what they said, quite strongly

> Planners are not very concerned about the current practices of using any kind of biomass for cooking ... (they) are guilty of neglecting the daily hard labour and health hazards run by a very large section of rural women of India (150 million). The State cannot accept this situation as given.... Missing in the Approach paper to 12th plan 'inclusion' or importance of making a concerted effort to meet the everyday needs of poor women, either in the context of reducing the drudgery associated with collecting firewood or preventing the pollution or health hazards associated with using them for cooking the food that feeds the households.[4]

However, these economists need to make a much stronger public demonstration of their clarion call, so that it does not remain hidden away in some little read document. Perhaps they and activists should start an advocacy movement.

Policy Imperatives

Effective strategies for accelerating the transition to cleaner and more efficient household energy among the 3 billion people in the world whose development prospects are held back by practices that affluent nations have long since left behind are, therefore, required as a matter of extreme urgency. This will require wider awareness of the problem, creation of a technical and regulatory framework, but most importantly, increased political commitment, and careful deployment of substantially greater financial resources than are currently allocated to this critical aspect of development. The basic public policy question facing the governments and the international community is whether the benefits of improved stoves are worth the financial investment that might be necessary. There cannot be any doubt that indeed this is so. In fact, it is also a moral imperative.

Public Health Dimension as a Key Driver to Policy Change

The linkage between household air pollution and non-communicable diseases (NCDs) is well established and the recent global push to tackle NCD burden could also serve as a key driver to bringing about policy change in use of clean cookstoves. The 66th World Health Assembly endorsed the WHO Global Action Plan for the Prevention and Control of NCDs 2013–20 (resolution WHA 66.10). The global action plan provides a road map and a menu of policy options for member states, WHO, other UN organisations and intergovernmental organisations, NGOs and the private sector, etc., on how best to bring about a 25 per cent relative reduction in premature mortality from NCDs by 2025.[5]

In its 66th Session, held in September 2013, the WHO South East Asia Regional Committee adopted an additional voluntary target of reduction of exposure to household air pollution, in addition to the nine global targets, that reads as follows '50% relative reduction in the proportion of households using solid fuels (wood, crop residue, dried dung, coal and charcoal) as the primary source of cooking' between 2010 and 2025.[6] Their Technical Working Group felt that since a baseline is available, progress can be easily measured.

India has adopted an additional target of 50 per cent reduction in indoor air pollution exposure by 2025 as part of its NCD monitoring framework and set up an 'inter-ministerial coordination committee on NCDs'. There is an urgent need to bring the cookstove agenda into these discussions.

Extent of Global Burden

We have noted in Chapter 1 the IEA estimates that there are currently about 2.7 billion people in developing countries who rely on cooking primarily on biomass. The worrying part is that this estimate is higher than before. People have been driven back to use biomass because of population growth, particularly amongst rural poor, rising liquid fuel costs and the global economic recession. About 82 per cent of these live in rural areas, although in sub-Saharan Africa, nearly 60 per cent of people living in urban areas also use biomass for cooking. The share of the population relying on the traditional use of biomass is highest in this area and India (IEA, 2010). The expectation is that ordinarily these numbers may not change much. IEA estimates in fact that if it is business as usual, this number may actually increase to 3 billion by 2030. For change to happen, and impact extreme poverty in any worthwhile manner, the enormity of the task is reflected in the assumption that to achieve this MDG goal by 2015, 1 billion people would need to gain access to clean cooking facilities including LPG stoves, advanced biomass cookstoves and biogas systems in this short time. Clearly this is not going to happen. It is also apparent that since these alarming statistics appeared, lots of discussion has been generated, but it is still mostly business as usual. So the danger of number of people globally increasing is quite real.

Status in India

According to the 50th NSS round conducted in 1993–94 and 55th round conducted in 1999–2000, 90.5 and 86 per cent, respectively, of all rural households in India used solid fuels as their primary cooking fuels. LPG penetration was not much but increasing. It would appear that things have not changed very much over the decade. NSSO report on Household Consumption Expenditure (66th round 2009–10) showed that 87 per cent of rural households used firewood and wood chips while 41 per cent used cow dung cakes. The percentage, according to Census of 2011, also virtually remains the same at 85.7 per cent. Firewood, crop residue and cow dung cake (all biomass fuels) have a penetration of 62.5, 12.3 and 10.9 per cent, respectively. The use of dung has increased from 10.4 to 10.9 per cent. LPG penetration in rural households has risen to 11.4 per cent from 5.4 per cent in 1999–2000 while that of kerosene has declined from 2.7 to 0.7 per cent (Figure 7.3). Biomass remains the most predominant source of cooking energy in rural areas with its use being uniformly distributed across all expenditure classes, with the richest 20 per cent moving up the energy ladder by using LPG. Those households who own cattle can also move towards biogas. In urban areas LPG has now become the major source of cooking energy, and should replace

Figure 7.3
Households' main source of cooking

(a) Rural households (millions)

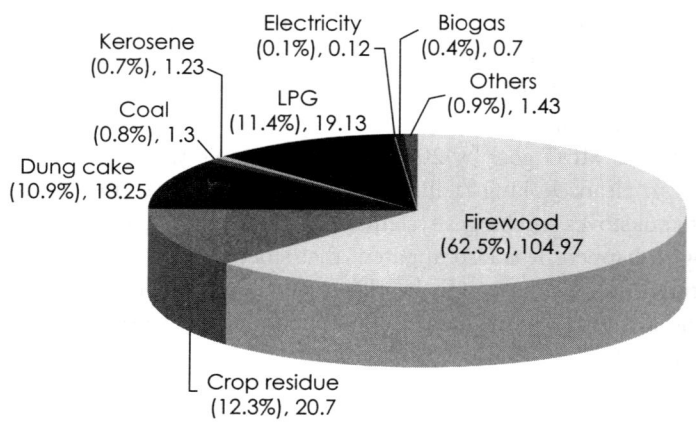

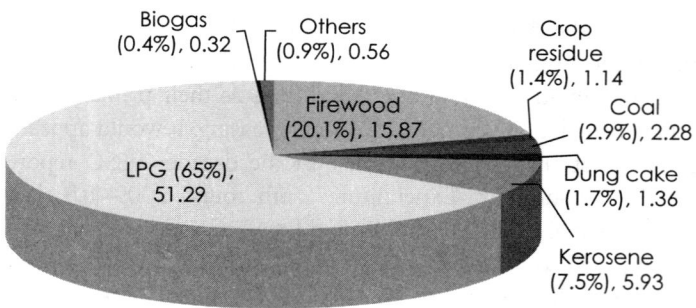

Source: Census of India, 2011.
Note: Figures in parenthesis are the percentages.

kerosene and wood. This would also be clearer from the state wise position which partly reflects the position of urban areas (Table 7.2).

These figures are in percentage terms. In reality, the number of households using biomass has significantly increased because of huge increase in population. The position has, therefore, actually considerably worsened in terms of numbers. The number of poor households

Table 7.2
State-wise distribution of households by type of fuel used for cooking (per cent)

State/Union Territory	Firewood	Crop Residue	Cow Dung Cake	Coal, Lignite, Charcoal	Kerosene	LPG	Electricity	Biogas
India	49	8.9	8	1.5	2.9	28.6	0.1	0.4
J&K	58.9	2.5	4.2	0	1.3	31.6	0.4	0.8
Himachal Pradesh	57.5	1.1	0.2	0	2.1	38.6	0.2	0.1
Punjab	13.4	6.5	20.4	0.2	3.2	54.5	0	1.4
Chandigarh	4.6	0.3	0.2	0.1	21.9	71.6	0	0.1
Uttarakhand	48.7	1.3	3.2	0.1	1.8	44.2	0	0.5
Haryana	26.1	14.1	14.2	0.1	1	44	0	0.3
NCT of Delhi	3.4	0.3	0.6	0.1	5.3	89.9	0	0.1
Rajasthan	61.8	11	3	0.1	0.9	22.8	0	0.1
Uttar Pradesh	47.7	8.7	23.1	0.3	0.7	18.9	0.1	0.2
Bihar	34.7	32.5	21.7	1	0.3	8.1	0.1	0.3
Sikkim	52.5	0.6	0.2	0.1	4.4	41.3	0.3	0.1
Arunachal Pradesh	68.7	0.7	0.1	0	0.7	29.2	0.1	0.1
Nagaland	77.9	0.8	0.1	0	0.6	20	0.1	0.1
Manipur	65.7	1.1	0.2	2.1	0.2	29.7	0.1	0.2
Mizoram	44.5	0.3	0.1	0.4	1.8	52.6	0.2	0.1
Tripura	80.5	0.8	0.1	0.1	0.6	17.6	0	0.1
Meghalaya	79	0.9	0.3	2.3	3.7	11.9	1.5	0.2
Assam	72.1	6.4	0.9	0.1	0.6	19	0.1	0.1
West Bengal	33.1	25.6	10	7.9	2.1	18	0.1	0.3
Jharkhand	57.6	4	7.2	18.1	0.2	11.7	0.3	0.1
Odisha	65	10.2	9.4	1.6	1.1	9.8	0.4	0.2

(Table 7.2 Continued)

(Table 7.2 Continued)

State/Union Territory	Firewood	Crop Residue	Cow Dung Cake	Coal, Lignite, Charcoal	Kerosene	LPG	Electricity	Biogas
Chhattisgarh	80.8	0.9	3.7	2.3	0.5	11.2	0.1	0.2
Madhya Pradesh	66.4	5.6	7.7	0.2	1.3	18.2	0.1	0.4
Gujarat	44	5.7	2.6	0.5	7.6	38.3	0	0.9
Daman and Diu	10.8	1.5	0.2	0.2	30.8	53	0.1	0.9
Dadra and Nagar Haveli	40.4	0.4	0.2	0.1	17.8	39.8	0	0.4
Maharashtra	42.6	4.5	1.2	0.2	6.5	43.4	0.1	0.7
Andhra Pradesh	56.8	1.4	0.6	0.3	3.9	35.8	0.1	0.7
Karnataka	57.5	2.9	0.2	0.1	5.4	32.5	0.1	0.9
Goa	20.7	0.9	0.2	0.1	4.1	72.7	0.1	0.4
Lakshadweep	54.8	10.7	0.1	0.1	13.7	16.6	1.2	0.2
Kerala	61.9	0.8	0.1	0.1	0.4	35.8	0	0.6
Tamil Nadu	43.5	0.6	0.2	0.1	6.9	47.9	0.1	0.3
Puducherry	18	0.3	0.1	0	10.3	70.5	0.1	0.1
A & N Islands	33.8	0.4	0	0	19.8	44.5	0	0

Source: Census of India, 2011.

using biomass is also then not likely to change much or actually increase. There is another factor to be kept in mind. In general, poorer households would be those having more individuals within each household and likelihood of exposure more because of the size/construction of the house. Therefore, the number of persons actually using biomass cooking or affected by it would be more than what is suggested by a percentage of households and the threat graver from the point of view of public health. Biomass cooking problem, therefore, is clearly that of the poor and, amongst them, the poorest.

Resource Availability of Fuels

Biomass

Estimates of biomass consumption remain highly variable since most biomass is not transacted on the market. There are varying estimates about biomass utilisation in the country. Biomass potential could be estimated from three major distinct sources, viz., wood, agro residue and cattle dung. The annual growth of woody biomass in the country has been estimated at 307.4 million tonnes. While forests are estimated to contribute 138 million tonnes, the social forestry, village forests, wastelands and roadside plantations together contribute the rest of 169.4 million tonnes.[7] As wood has multiple uses, only a fraction (NSS estimate suggest 176 MT) of the same is available for energy purposes. An area of about 107 million hectares has been estimated to be degraded with 64 million hectares categorised as wasteland, which includes degraded forests.[8] This incidentally shows that there is theoretically substantial scope for dedicated energy plantations, which is another policy requirement, especially with now known fast growing species. Further, around 150 million tonnes of surplus agrobiomass resources are generated every year in the country. About half of the surplus residues are burnt in the fields. Therefore, it could be safely presumed that around 70 million tonnes of agro-residues are available.[9]

Biogas

Another bio-resource in India is cattle dung. Country has a livestock potential of around 280 million. With a very conservative figure of 5 kg dung per cattle per day, this cattle population produces around 500 million tons of wet dung annually. With broad estimate of 1 cubic metre (m^3) biogas per 25 kg wet dung, this dung is sufficient to produce around 20 billion cubic metres biogas per year. Theoretically it would be possible to set up 27 million biogas plants of 2 m^3 capacity. However, due to practical limitations such as availability of water, cattle holding per household, absence of stall feeding with cow dung being dropped in fields, and use of cow dung cakes for cooking, etc., Ministry of New and Renewable Energy, Government of India, has estimated biogas plant potential to be 12 million plants. Depending on the family size, this would mean a regular supply of fuel for 60 to 85 million people. However, in practice, we may be able to reach only almost 50 million people, which will also be a good achievement. In addition, Biogas could also be generated from other organic resources such as sewage, municipal solid waste and distilleries.

Kerosene

Around 0.7 per cent of rural and 7.5 per cent of urban households use kerosene for cooking. Kerosene has the advantage of being efficient, quick and easily controllable, which are all convenient features as compared to other rural cooking devices. However, kerosene stoves give off an unpleasant smell and their safety is questionable. Even though government provides very high quantum of subsidy for PDS kerosene it is still a costly fuel. Further, the total amount available per family is so small that it simply cannot sustain cooking needs. For these reasons, as also easy availability of free biomass, and widespread use of kerosene for lighting, its use for cooking purposes in rural areas will remain only a very marginal option.

Liquid Petroleum Gas

The marketing of LPG in India started way back in 1955–56. LPG use for cooking purposes has witnessed major growth during last 5 years. Its penetration has significantly increased and now 11.4 per cent rural households (19.13 million) use LPG for meeting cooking energy requirements. There is a school of thought that believes in spreading LPG to rural areas in a big way, and of course, this would be justified. However, it is apparent that the growth of LPG in rural areas will be slow because of the many difficulties involved in making this happen which we shall discuss subsequently, and the focus could, and perhaps should be, to cover still uncovered urban areas or groups that are much easier to tackle. Delhi has started a programme to make the City State kerosene free. Why should Chandigarh have so many households using kerosene?

National Efforts

The past 20–30 years have seen many diverse programmes of household cooking energy across the globe, from small and medium-scale NGO and community-led initiatives to pilots with help of donors or

Access to Cooking Energy 181

multilateral organisations and ambitious national programmes. The largest programmes were in China and India during the 1980s and 1990s, both roughly at the same time. It is important to note that these were in response to the grave perceived threat at that time of disappearing forest land. They were, therefore, not designed specifically to achieve low pollution exposures. There were also programmes for fuel wood plantations to provide the necessary fuel. In the early programmes it was commonly assumed that the benefits of the improved stoves were so self-evident that an initial intervention would lead seamlessly to a self-sustaining programme. This was not to be. In fact, failure of past fuel wood plantations and improved stove programmes generally created pessimism in the development community about the relevance and effectiveness of interventions on household energy access. This contributed significantly to the lack of policy attention to this area and consequently the allocated resources for intervention (Barnes et al., 2012). Perhaps it has also led to the feeling that it is too difficult a problem to resolve, and LPG, if and when it is possible to do so, would be the best alternative.

Policy Landscape (1980–2013)

1980–2005

In the 1980s, there appeared to be some concern to address the problem of rural cooking energy. In 1981, the Government of India launched the National Project on Biogas Development (NDBP). This was renamed in 2002–03 as the National Biogas and Manure Management Programme (NBMMP) with the twin objectives of meeting cooking energy needs of the rural areas and producing enriched manure. It was also recognised that it would help village sanitation and hygiene, reduce drudgery of fuel collection for women, reduce consumption of fossil fuels as well as fuel wood and reduce methane emissions. This programme continues till today but is being implemented in a somewhat routine manner. India became a pioneer in family-sized biogas plant technology. In 1983, the Government of India also started the National Programme on Improved Cookstoves (NPIC) with multiple objectives of conserving wood as fuel, reducing smoke emissions and diverting them out of the house through chimneys and improving health. India has been one of the few countries with a sustained commitment to promoting improved cookstoves, although this has really not been followed through to its logical course.

This programme lasted for two decades but was given up in 2002 as a central sector programme and transferred to the states for funding and implementation. The programme thereafter withered away except for a few local initiatives here and there. During this period, some efforts were also made for dissemination of box solar cookers with central subsidy, which were also soon dispensed with. There have been some developments related to solar dish cooking technology. The focus then shifted to spread of LPG in rural areas but which over the years has got only marginal success. In general, since the start of this century, rural cooking energy has taken a back seat.

Integrated Energy Policy, 2006

This was also subsequently reflected in the discussion on this issue in the Integrated Energy Policy Report which came out in 2006. There was almost a cursory reference to this matter stating that because of the various adverse effects the urgency to meet the challenge should be high. The solution, almost resurrected from the past, was more energy and fuel wood plantations, in spite of them having generally been a failure and understandably not found to be a long-term solution, and looking into the future, supply of subsidised LPG to almost 40 per cent of the rural population. It was suggested that a goal may be set for supply of LPG, NG, biogas or kerosene to all within 10 years—none of which was even remotely possible. A complicated system of disbursal of subsidy was also suggested. Community-sized biogas plants were recommended, even though the difficulty of voluntary collaboration was recognised. By that time such plants had mostly failed. There was, therefore, no real analysis of the problem and no practical recommendations.

2006–13

The cookstove programme was attempted to be resurrected through the National Improved Cookstove Initiative in 2009, but it has made only faltering progress although there is a programme of a few million. A biogas programme is running with somewhat reducing annual targets and LPG penetration is not substantially increasing in rural areas. As we have noted earlier, there is also no significant emphasis on this matter in the 12th Plan. The expert group studying the low carbon strategies merely states that 'inclusive growth' means all households have access to clean cooking fuels such as natural gas or LPG making a 'secular shift' from traditional biomass to modern commercial energy. The how or the feasibility or the costs have not been examined. In a separate reference to increase forest cover it talks of the need for more efficient biomass stoves so that pressure on firewood for cooking is reduced.[10] The time has come to change this, and to do so urgently.

Programmes in India

National Programme on Improved Cookstoves

Initiation

The main objectives of NPIC were reducing deforestation, limiting the drudgery of women and children, reducing cooking time and improving employment opportunities for the rural poor. Fuel efficiency was the main criterion as the full benefits of clean combustion were not well understood and stove technologies to achieve it were not well developed (Venkataraman et al., 2010). It concentrated on the poorest rural households. The newly created independent Ministry of Non-Conventional Energy Sources was responsible for planning, funding, setting targets and approving stove designs. However, it was essentially a state-led programme with state-level agencies implementing it in decentralised mode through local government agencies or NGOs. A technical backup unit (TBU) in each state trained rural women or unemployed youths to become self-employed workers to construct and install stoves. Many also conducted laboratory research to develop and test stoves.

Progress and Closure

Between 1984 and 2002, the programme distributed more than 35.2 million improved stoves against an estimated potential of 120 million in the country. Most improved stoves were built of mud and a few steel components, with a potential life of only two years. In a 1996 survey, it was found that 60 per cent of existing improved stoves were in use in households but that the improved stoves constituted only 7 per cent of the total stock of stoves in the rural areas and that nearly 60 per cent of households did not wish to change their traditional stoves. In 2000–01, the focus of the programme was shifted towards the promotion of durable fixed-type improved cookstove with chimneys, with minimum 5-year lifespan and with a view to improve indoor air quality in rural kitchens. Six new models of durable cookstoves with chimney were developed in 2001–02. In addition seven models developed and field tested in 2000–01 were also promoted along with the other 15 approved models.[11] In 2003, Planning Commission transferred the programme with full implementation authority and funding support to the states. Some states continued in a fragmented manner and others closed down the program completely. In a later survey, it was found that only 5 per cent of the rural households had improved stoves (Zhang et al., 2006).

Assessment

Despite the good potential, the programme was not able to meet its objectives and targets.[12] The low adoption rates under the programme indicated certain problems identified by evaluations and analysis.[13]

- Models were designed so that even the poorest customers could afford them. Valued at about US$5 or less, these stoves represented an improvement over a three-stone open fire; nonetheless they were rudimentary devices. They were typically built on-site using inexpensive materials.
- Among those that were adopted, poor quality and lack of maintenance resulted in a lifespan of two years at most and typically much less. Most stoves used inexpensive materials with low durability, leading to poor performance and short lifetimes.
- There was inadequate emphasis on commercialisation and after sales servicing seen as crucial for effective and sustainable uptake. Large subsidies deterred the development of markets.
- There was insufficient interaction with users, self-employed workers and NGOs. Designs did not meet household needs and there was very poor acceptance of user training.
- In some cases models were pre designed—this may not have suited a family's needs. It also prevented competition.
- Quality control, monitoring, and evaluation for installation and maintenance of the stove and its appropriate use were lacking.
- Persistent problems with chimneys, including their covers, caused most users to revert to traditional stoves.

- High levels of subsidy (approximately 50 per cent of stove cost, and in some states more) were found to reduce household motivation to use and maintain the stove. Generally the households which got the largest subsidies showed the poorest maintenance. In such cases, the life was only about 6 months. A dependency syndrome was also created.
- Lack of a large programme of awareness generation.
- Too much emphasis on setting and achieving targets.
- Unlike the Chinese example, non-certified artisans were able to install stoves; stove parts were purchased from non-approved suppliers and people often themselves made modifications to suit their cooking preferences. Thus, the improved stoves lost their effectiveness.

Improved Cookstoves during Intervening Years

It is indeed sad that the programme was virtually abandoned instead of utilising the immense learning experiences to develop a more comprehensive and sustainable programme. Valuable expertise and skills built were lost. In our view this was a big mistake, which has yet to be rectified. Probably the thinking was, and still largely is, like the faith in the centralised grid for electricity access, that the solution lay only in the widespread extension of LPG in rural areas. There could also be a feeling that cookstove programme would be difficult to implement.

The consequent absence of support for cookstoves since then spawned a number of state, private, or commercial activities, e.g., Appropriate Rural Technology Institute (ARTI) in Maharashtra, Self-Employed Women's Association (SEWA) in Gujarat and Technology Informatics Design Endeavour (TIDE) in Karnataka. There were small attempts at selling improved stoves in localised areas. There may be some support from CSR or Foundations. Envirofit started off as a private enterprise. BP also used the stove designed by the Indian Institute of Science, which was later transferred to a private company, First Energy. But this stove was designed to use pellets. They started subsidised sale of these stoves and the pellets which proved very popular giving the impression that this was maybe a breakthrough. However, as soon as prices of the pellets were raised the consumers declined substantially, thus discounting the hopes of the success of this model. This model has of course been successful in commercial stoves which are now picking up in a big way in market mode.[14] However, in general, the progress has been agonisingly slow for household stoves.

Improved Cookstove Initiative

Initiation

Encouraged by Mr Shyam Saran, then Climate Change Envoy to the Prime Minister, and with the help of a few dedicated stakeholders, MNRE started a new initiative on cookstoves in December 2009.[15] The very ambitious goal was 'to achieve the quality of energy services from cookstoves comparable to that from other clean energy sources such as LPG' through the 'development of next generation household cookstoves, biomass processing technologies and deployment models' (MNRE). The experts understood that the major premise behind the NCI was that the combination of newly understood benefits and development of new technologies (for high combustion efficiency and low emissions) provided a renewed rationale and new possibilities for this goal (Venkataraman et al., 2010).

Content

The following measures were thought of. First, make standards which would set minimum quality and performance requirements. Second, set up National Centres of Excellence that could test such stoves and establish testing protocols. Third, develop different stoves to meet these standards. Fourth, pilot test them through deployment which would show their acceptance, etc. Fifth, pilot deployment models, including financing arrangements. Sixth, look into manufacturing arrangements. Finally, launch a bigger programme, which could then be up-scaled into a full national programme.

In order to maintain quality, cookstove models would be tested as per stipulated test procedures to conform to required performance. Three Biomass Cookstove Test Centres one each at IIT, New Delhi; Institute of Minerals & Materials Technology (IMMT), Bhubaneswar; and University of Agriculture & Technology, Udaipur, Rajasthan, have been funded by MNRE to cater to these requirements for different zones in the country. Standards have since been improved and revised through Bureau of Indian Standards (BIS). These are given in Table 7.3. It may be noted that such standards have still not been prepared internationally which is a task which needs to be urgently taken up.

Table 7.3
Standard performance parameters for cookstoves

S. No.	Type of Biomass Cookstove	Standard Performance Parameters		
		Thermal Efficiency (%)	CO (g/MJd)	PM (mg/MJd)
1.	Natural Draft Type	Not less than 25	≤ 5	≤ 350
2.	Forced Draft Type	Not less than 35	≤ 5	≤ 150

Source: Ministry of New and Renewable Energy.

Natural draft and forced draft models of single part metallic and metal plate ceramic (ceramic composite/industrial insulating materials) cookstoves have been developed and are being manufactured in the country. The MNRE has so far approved six models each of natural draft and forced draft cookstoves for domestic size and three designs of Forced Draft for community size. However, while these stoves show improved performance, their durability is a question mark. Durability must be increased. Costs would also need to be kept under control. A national programme would need to take care of both these aspects. Costs would decline if the number to be distributed is large.

Box 7.2 Improved cookstove in China

Let us also see briefly the experience of China. The world's largest publicly financed improved stove programme was implemented in China. Started also in the early 1980s, 185 million of China's 236 million rural households had improved biomass or coal stoves by 1998. The programme differed substantially from that of India, though operating over the same period. While the Indian programme focussed on low-cost stoves for the poorest households, this was largely across the population. An advantage was that although rural Chinese populations were/are poor, they do have greater effective purchasing power than in many developing countries, allowing the majority of consumers to purchase stoves at close to full cost. The subsidy element was 26 per cent. Among the key features of the Chinese programme reported to have contributed to its success were decentralisation of administration, a commercialisation strategy that provided subsidies to rural energy enterprise development, and quality control through the central production of critical components (such as combustion chamber parts), which made household

(Box 7.2 Continued)

(Box 7.2 Continued)

assembly easier and the product more robust, and engaging local technical institutions in modifying national stove designs to local needs. National-level stove competitions generated public and media interest, a bidding process among counties for contracts allowing the best-placed counties to proceed first; financial payments were provided to counties only after completion of an independent review of their achievements. No large flows of funds came from the centre as the major financial contributions were provided by local governments. The Chinese programme shifted norms—most biomass stoves now available on the market have flues and other technical features that classify them as improved. Early on, the technical design work was done in a national design stoves training and design centre. To lower stove costs while ensuring quality, the technical centre developed standardised inserts. Close association with stove retailers and cooperation with customers played a vital role during the stove design phase. Extensive consumer testing and marketing preceded large programme roll out. Stoves had to meet certain engineering standards in the lab to ensure smoke removal and energy conservation. Evaluations were held in every county to ensure that households used the stoves as intended. Only certified trained artisans assembled the stoves in houses, retailers and in cooperation with customers played a vital role during the stove design phase.

Source: The authors, as summarised from readings.

National Project on Biogas Development

Experience Gained

This programme was started in 1981 and continues till today. India has now the experience of installation of family-sized biogas plants for three decades. National Project on Biogas Development (NPBD) has a long trail of experimentation and innovative approaches addressing operational and technical glitches. It invested in design research to make biogas suitable to different climatic conditions and develop models that took into account local problems like low temperature, less availability of water, alternate feeding material and proper feedstock mix. Standardised models are now available for individual and community-sized plants with optimisation of biogas yield, cost and durability. Extensive infrastructure for training exists and a large pool of skilled manpower in rural areas has been developed for taking up construction and maintenance of such plants.

Progress

By March 2013, around 4.5 million family type biogas plants had been set up in the country which is about 39 per cent of the estimated potential. State-wise potential and achievement up to 31 March 2013 is given in Table 7.4.

Table 7.4
Family type biogas plants: State-wise estimated potential and cumulative

State/Union Territory	Estimated Potential	Cumulative Achievements as on 31 March 2013
Andhra Pradesh	1,065,000	504,887
Arunachal Pradesh	7,500	3,382
Assam	307,000	101,905
Bihar	733,000	129,523
Chhattisgarh	400,000	44,594
Delhi	12,900	681
Goa	8,000	4,039
Gujarat	554,000	426,074
Haryana	300,000	58,502
Himachal Pradesh	125,000	46,889
J&K	128,000	2,945
Jharkhand	100,000	6,746
Karnataka	680,000	457,571
Kerala	150,000	136,967
Madhya Pradesh	1,491,000	335,648
Maharashtra	897,000	840,036
Manipur	38,000	2,128
Meghalaya	24,000	9,496
Mizoram	5,000	4,520
Nagaland	6,700	7,149

(Table 7.4 Continued)

(Table 7.4 Continued)

State/Union Territory	Estimated Potential	Cumulative Achievements as on 31 March 2013
Odisha	605,000	260,056
Punjab	411,000	153,162
Puducherry	4,300	578
Rajasthan	915,000	68,341
Sikkim	7,300	8,474
Tamil Nadu	615,000	220,251
Tripura	28,000	3,218
Uttar Pradesh	1,938,000	434,139
Uttarakhand	83,000	15,804
West Bengal	695,000	364,638
A&N Islands	2,200	137
Chandigarh	1,400	97
Dadra and Nagar Haveli	2,000	169
KVIC		17,637
Total	12,339,300	4,670,383

Source: Ministry of New and Renewable Energy.

Approach

Over the years the annual targets were in the range of around 1–1.5 lakh biogas plants per year. During the 10th Plan period, 5.63 lakh plants were installed. During the two years, 2007–09, 1.97 lakh plants were installed. In 2009 the subsidy per plant was raised to ₹8,000 per plant and the annual targets were raised to 1.5 lakh. This led to an increase in the installation of biogas plants as reflected in the performance during the 11th Plan period as given in Table 7.5.

For the 12th Plan, however, the target is 5.75 lakh plants with the annual target for the last 3 years being revised downwards to 1.10 lakh. This means that the momentum gained would be lost to a certain degree. Actually the aim should have been to go to 2 lakh per annum. The average subsidy, however, has been further raised to ₹12,000 per plant.

Table 7.5
Year-wise achievement for family type biogas plants during the 11th Five Year Plan

	Target (in Lakh)	Achievement (in Lakh)
2007–08	1.04	0.39
2008–09	1.24	1.08
2009–10	1.50	1.20
2010–11	1.50	1.51
2011–12	1.50	1.80
Total	6.78	6.48

There would, evidently, and unfortunately, be a budgetary constraint which will restrict numbers.

There is a need to structure and implement this programme better. Earlier, targets were distributed pro rata among the states, and further downwards similarly amongst the districts and blocks. In 2009–10, targets were raised or reduced as per the request of individual states or the interest they showed. It was also suggested to them that these should not be distributed among districts, blocks and villages in a pro rata manner—rather there should be a cluster approach to achieve volumes in certain areas where potential and acceptance is good. Considering that in technology dissemination volumes are of utmost importance, for creating an effective operation, repair and maintenance services, the approach of allocating token targets which translates into around one plant in every six villages neither creates an environment for a proper institutional framework nor helps in exploiting the potential in foreseeable time frame. This must be given up.

One of the major problems of the programme is the total dependence upon the annual budgetary resources, which is perhaps getting reflected again in the reduced targets. Whatever may have been written in the Plan documents, annual targets are fixed at the Government of India level from year to year and this is subject to resource availability in a particular year. A 5-year target should be prepared. Funds should be ensured and guaranteed and allowed to be carried over from year to year. Annual budgetary restrictions limit the ambition because we need to have higher targets. The uncertainty of fund availability also prevents the development of a momentum which is necessary to go for higher numbers, as has become evident from the numbers given above. If funds become available many exciting initiatives can be taken forward.

Functionality of Plants

Although there are various claims about success and criticisms of failure, the National Census in 2011 has revealed that around 1.1 million households were meeting their cooking energy needs mainly through biogas. This may give a perception of poor functionality, but the report also added that others were being used in combination with other fuels. Independent evaluations of the programme carried out over different years suggest that this number is actually much higher. In general, those plants, which were installed in the earlier years, may not be working, partly because of many of them passing their 15-year periods of longevity, and cases of lack of maintenance by the beneficiary or the system. The problem has been more acute in some states such as Uttar Pradesh, Bihar and Haryana where failure rates would be more and where progress subsequently has been poor. The more recent plants have generally found to be functioning better.[16] M/s Andhra Pradesh Industrial & Technical Consultancy Ltd has evaluated plants set up during the 11th Plan. The study found around 95 per cent plants functional.[17] The real challenge in the future will also be to tackle the state of Uttar Pradesh and Bihar, where there is huge unexploited potential.

The Public Accounts Committee examined the National Project on Biogas Development. In its 96th report laid in Lok Sabha on 30 April 1987, many pertinent observations were made which are reproduced below

> ... the Committee have pointed out that the programme has become widely acceptable and very popular with marginal farmers, village women, adivasis and other economically backward people as it is convenient and economical, has multiple benefits and is the first major step towards improving rural sanitation and environmental hygiene. The setting up of more plants will reduce ecological disturbances and deforestation. Biogas is probably the only programme capable of relieving the miseries and drudgery of village life, especially of village women.... Besides, biogas produces good quality manure from organic waste materials like cattle dung which is

rich in humus and micro nutrients and provides nitrogen phosphates and potash to crops. The Committee hoped that Government especially the Planning Commission, should take note of the economic and employment potential of the biogas projects vis-à-vis chemical based fertiliser plants and earmark suitable amount for development and propagation of biogas projects in the country.

Focus on LPG

Vision 2015

Launch of RGGLVY

In October 2009, the Ministry of Petroleum and Natural Gas finalised the Vision 2015 for the oil sector for 'Customer Satisfaction and Beyond'[18] and launched the Rajiv Gandhi Gramin LPG Vitaran Yojana (RGGLVY). This envisaged setting up of small-size, low-cost LPG distributorships in rural areas for release of cooking gas connections to people in villages.[19] It was targeted to increase the number of LPG households by the year 2015 such that the population covered under LPG would be about 75 per cent, including the aim of 100 per cent LPG coverage in all the towns with population 5 lakh and above. It was planned to release most of the new connections in the rural areas amounting to 55 million, where LPG coverage is still low, so as to shift long time kerosene users to LPG.

Main Features of the Scheme

Main features of the scheme are summarised as under

  i. The agencies under the RGGLV will be of small size requiring lesser finance/infrastructure.
 ii. The agencies would penetrate deeper into the rural areas. RGGLV distributors may be viable for around 1,500 customers in the cluster of villages being served.
iii. The distributor will be able to recover the capital expenditure by the time 1,800 new LPG connections are released. The indicative net income of the distributor would be about ₹7,500 per month.
 iv. In order to ensure that growth of LPG usage is evenly spread, OMCs will be assessing/identifying locations under RGGLV in

rural/low-potential areas in all states/Union Territories so as to achieve target growth of at least 50 per cent LPG population coverage in each district.

v. In addition, small size LPG cylinders of 5 kg capacity were launched to extend LPG reach to hilly terrain and interior areas on account of convenience in transportation.[20]

However, against the overall aim of 55 million new LPG connections between 2009 and 2015, especially in rural areas, as on 1 November 2012, Public Sector Oil Marketing Companies could enrol only 2.1 million domestic LPG customers through these RGGLV distributorships in the country.[21]

Subsidy Support

The vision also envisaged that support should be provided to poor households to switch over from kerosene to LPG by leveraging the corporate social responsibility fund of oil PSUs along with the state governments. The Expert Group on 'A Viable and Sustainable System of Pricing of Petroleum Products' constituted by the Government of India in February 2010 made the recommendation that 'as a clean cooking fuel, LPG is a merit good and subsidy to poor households may be needed and justified. The level of subsidy should be fixed by the government on the basis of ability to pay, and should be paid out directly from the Budget'.[22]

Brief Analysis

There were some difficulties with this approach. First, there was no calculation of actual subsidy per cylinder or annually per household and, consequently, the additional annual subsidy required to supply LPG to additional 55 million households. Second, would this be for urban or rural households, or both? Would it be possible to provide this from the budget? Perhaps the fiscal deficit problems of recent years combined with rising oil price and a weakening rupee will bring in more realism, which indeed is slowly creeping in. Third, there was an unrealistic assumption that CSR would be able to supply free connections, leave alone the fact whether it is desirable to do so. Fourth, there also appeared to be a presumption to shift rural households away from kerosene to LPG, which is erroneous. It is well known that very little kerosene is being used by rural households for cooking which is also declining. There are other difficulties too in a shift to LPG which we shall discuss a little later. No wonder that progress in promotion of LPG in rural areas has not been very much. During 12th Five Year Plan period, 102.5 million LPG connections are planned.[23] However, the trends suggest that this figure may be too ambitious and achievements will most likely be in urban and peri-urban areas.

Solar Cooking

We may also briefly consider this area which has become a somewhat mature and cost-competitive technology for meeting partial cooking needs. Solar cooking devices are now available in different sizes, technologies and models for domestic as well as large applications, where the future appears to be brighter.

India has also been a pioneer in using two types of solar concentrating technologies for the purpose of steam generation for various applications. One is based on fixed receiver East-West automatically tracked parabolic concentrators (Scheffler dishes) and the other on fully tracked receiver and Fresnel reflectors (Arun Technology). Systems based on fixed receiver technology are under installation for many years and about 80 systems covering over 22,000 m^2 of dish area have been installed so far. The second technology is under pilot scale demonstration. Systems based on these technologies have been installed for cooking, laundry and process heat applications. The world's largest system for cooking in community kitchen was installed at Shirdi in Maharashtra in 2010 to cook food for 20,000 people daily and is saving around 60,000 kg of LPG annually. These are for larger applications for which there is a great scope in this country.

Dish solar cooker is a device which focuses solar energy at a point in the centre of solar dish and can cook all kinds of food using pressure cooker. Cookers for cooking food for about 10–40 people are available in different sizes. The most commonly available cooker is of 1.4 m diameter to cook food for about 10–15 people. These cookers have been found to be popular in Uttarakhand both for individuals and mid-day meals in schools. Over 2,800 dish cookers have been installed so far in the state during the last 7 years. Field evaluation of these cookers has revealed that cookers were in use even when the food serving time in summer is between 10 and 10.30 a.m. and between 12 and 12.30 p.m. in winter. Under the field conditions if used to cook one time meal in its full capacity for 300 days a year, one dish cooker could save 50 kg LPG and reduce LPG consumption by 3.46 standard 14.2 kg cylinders. On normative basis, it would amount to a saving of ₹1,050 per year @ ₹21 LPG per kg. The problem was that the solar cooker could be used only for one meal and, therefore, could not be utilised to its proper potential for mid-day meals.

There are smaller cookers also. Up to March 2013, around 6.5 lakh box-type solar cookers and 12,000 dish-type solar cookers have been deployed in the country to meet the cooking energy needs. The box-type cookers have been used since long but they are no longer very popular. The ease of cooking gas does not encourage people to go for this. It takes time, not everything can be cooked and the box has to be shifted from time to time. It has hardly spread to rural areas and it does not appear that it will do so. In more recent years, the dish-type cooker was also devised. This costs around ₹6,500 for a standard 1.4 m² size. This could cook on a normal sunny day for 300 days in a year. This has great potential for areas where biomass collection is a problem and the sun is good as in Ladakh. Under the Ladakh Renewable Energy project, there was a provision for 4,500 cookers at a subsidy of ₹3,000/m² of solar dish cookers.[24] Of these, around 2,000 dish cookers have been distributed.[25] A proper evaluation needs to be done. These could be very useful cooking devices for areas where biomass is difficult to collect or is virtually not available; it is very costly to supply diesel or kerosene and the solar radiance is very high. Indian army and para-military border posts would be ideal locations too.

Lessons Learned

A number of studies globally have looked into household cooking energy and related issues. After reviewing the experience of household energy projects globally, and their success and failure factors, the World Bank suggests the following important lessons (IEA, 2010). These are discussed below in the context of a possible programme:

- A holistic approach to household energy issues is necessary looking at their contribution to social transformation and poverty reduction.

Access to Cooking Energy   197

- A behavioural shift from traditional ways of cooking used over generations to a new technology is not going to be easy. Therefore, strong and innovative public awareness education and information campaigns regarding the health, environmental, gender and economic benefits are critical to stimulate demand and are essential pre-requisites for successful interventions. Households need to be sensitised to the risks they incur by cooking with inefficient stoves and the benefits of shifting to better stoves. Programmes that have assumed that households would adopt spontaneously improved stoves or participate in forest management initiatives have failed. Households must become convinced. Only this will cause the 'pull' effect as against traditional 'push' programmes. However, such a large programme can only be undertaken after the technical framework has been developed.
- Active participation of communities, governments, NGOs and the private sector is fundamental. For example, local communities need to be involved at an early stage to ensure that they own supply-side forest management initiatives. They can also do the advocacy for stoves better, especially if they are women's self-help groups (SHGs).
- Successful programmes pay attention to the needs and preferences of the users of improved stoves. Targeting households susceptible to buying and using these improved stoves and working with them to supply a suitable stove that responds to their needs is critical. At first, this target group is usually not the poorest of the poor. By first focusing on households that can afford to adopt an improved stove, the programme can subsequently capitalise on the benefits of the demonstration effects produced. Experience has shown that when these factors are ignored, stove dissemination rates are low, and programmes are not sustainable. In India, we also have to recognise that there might be regional differences and preferences.
- Durability of improved stoves is important for their successful dissemination. For households that can afford an improved stove the decision to adopt one or not includes their perception of its durability. The durability depends on the quality of the materials used in the production of the stove, the resistance of the stove in the climatic context where it is used, how it is used, proper maintenance, etc. It is important to account for durability issues in the design and construction of improved stoves, in addition

to technical considerations, such as heat transfer efficiency and combustion efficiency. Ordinarily, an improved stove should have a life of 5 years.
- A market-based approach in the commercialisation of improved stoves is often viewed as the best way to ensure sustainability of programmes. This is based on the evidence that subsidised programmes do not continue when donor or public funding dries out. (This is actually an argument against small donor programmes started in isolation.) Evidence indicates, however, that a certain level of public funding is necessary at the initial programme stages for improved stoves programmes to take off (or for the market development to take place). This is particularly true in settings where the business environment is not well developed. Funding is usually needed to support R&D, marketing, quality control, training related to stove design, testing and maintenance and monitoring and evaluation. Work on developing stoves standards and certification protocols rely on the availability of public funding. Without this initial support, small enterprises find it difficult to participate in improved stoves programmes, and scaling up is unrealistic. A challenge is to determine what level of public funding is adequate and the timing to transition to a fully market-based business model.
- With microfinance, the poor can gradually afford an improved stove. Availability of improved stoves and cleaner fuels is one thing, whereas their affordability is another one. Efficient, safe and durable stoves may cost $30–40 or more. Cheaper ones may not give the required benefits or be durable. Programmes that have included microfinance options to help households afford the stoves tend to be more successful. The poor need to have a time horizon to gradually pay for the improved stoves. Therefore, the challenge of the high upfront cost has to be met. For example, in Bangladesh, Grameen Shakti has been working with international donors to provide cookstoves as part of its microfinance activities. This dimension is very important. Having an improved stove is not perceived as a first priority by the poor, but by integrating the adoption of an improved stove in a broader programme, creating opportunities to generate income is a different proposition. (Micro finance in India is limited, so different options for bank financing would need to be found.)
- Good models of stoves need to be built based upon robust engineering which should pass standardised criterion for construction and performance. This also needs development of a supply

Access to Cooking Energy 199

infrastructure for various components as was done in China. There would be need for decentralised production and ability to take these to rural markets in large numbers. These should be tested. But as the Global Alliance for cookstoves points out, the breadth of cookstoves required to meet consumers' needs and wants does not yet exist. There is also a desirability issue—the size, colour, shape, user friendliness, ability to cook locally preferred foods.

These lessons only serve to underline the many challenges that need to be addressed before we can have a large-scale cookstove programme.

New National Programme

We strongly recommend the resurrection of the improved stove initiative along with an accelerated biogas programme. The following steps could be taken in sequence:

- Finalise standards for stoves developed for various characteristics such as its combustion efficiency, carbon emissions and particulate matter. This has already been done. But standards for durability are also needed.
- After extensive testing, different models should be approved. There is a case for governmental assistance in this development.
- As a part of this process, national and regional testing and technical support centres should be set up, or strengthened, as may be required.
- Thereafter, there should be pilot projects at different places, which should be publicly assisted for allowing consumer preferences for different kinds of stoves. It has been suggested quite rightly that the areas to be selected for such roll-outs should be those where people are familiar with stoves and there is proven history of usage, where collection of biomass is becoming tougher or is being partly bought and where there is better scope for setting up a market infrastructure.
- If required there should be design changes and the stoves should finally be approved.
- There should now be intensive regional promotional programmes for public education and assistance in creation of a marketing infrastructure in those selected areas.

- Assistance should also be given for setting up this infrastructure as well as for intensive training to local people, ideally women, who could become the sales and support persons to the manufacturers.
- People should be given a voucher for purchase of such approved stoves which would be available in the local market with a limited subsidy perhaps of ₹1,000 per stove (this would need fine tuning). The prices of approved stoves would be fixed for the year.
- It would be essential to have either micro financing or financing through women's self-help groups or bank financing so that the costs of the stoves could be paid over time.
- Thereafter a national roll-out can be done. Reaching that stage could take 5 years.

The new initiative on cookstoves was conceived in somewhat this way in December 2009. Setting up of national test centres was approved. Standards, though delayed, have been finalised. Some stoves models have been cleared (Table 7.6).

GIZ has tested some stoves in Bihar, Uttar Pradesh and West Bengal. The results are not definitive in terms of acceptability. And doubts remain. There must be an assurance of performance. But these stoves have a problem of durability as well as cost. That will lead neither to scale nor to public acceptance. There is a developing feeling that not only should durability be more but that reduction of smoke and particulate matter emission should also be more. Discussion with some experts[26] suggests that there is perhaps now need to move further from an 'improved stove' to an 'ultra-clean stove' in order to achieve scale and make a real difference. It is suggested that such a stove must be the forced draft one. There are also efforts going on to develop a stove/s, which could provide its own power through use of thermo electric generation. There are also efforts to develop a stove which could use cow dung by using bio fuel technology. Two US companies are making these efforts and IIT Delhi is also experimenting. This could lead to a breakthrough and may actually be the real answer. But the other steps have to be taken starting with pilot testing at different places. This exercise must be supported liberally by the government. At the moment we have initiated efforts for getting some CSR support for the research and field testing of some new type of stoves as mentioned above. Earlier it was envisioned to roll out 100,000 stoves, which has not happened yet. Perhaps we need to wait a little more

Table 7.6
Approved models of cookstoves

S. No.	Models	Power Output
Natural Draft Cookstoves—Domestic Size		
1.	Harsha (Fuel-Wood) (CSIR, IMMT Design)	1.83 kW
2.	Bio-classic (Fuel-Wood)	1.49 kW
3.	Greenway Smart Cookstove (Fuel-Wood)	0.8 kW
4.	Firenzel (Fuel-Wood)	0.74 kW
5.	Adarsh (Nirmal) (Fuel-Wood)	0.89 kW
6.	Chulika (Fuel-Wood)	0.74 kW
Forced Draft Cookstoves—Domestic Size		
1.	Oorja (Fuel-Pellets)-IISc–Design	0.7 kW
2.	TERI SPT-0610 (Fuel-Wood)	1.08 kW
3.	Eco chulha—XXL (Fuel-Wood)	1.10 kW
4.	Agni Star (Fuel-Rice Husk)	2.16 kW
5.	Ojas (Fuel-Pellets)	1.99 kW
6.	RAMTARA (Fuel-Pellets)	1.0 kW
Forced Draft—Community Size		
1.	Eco chulha-XXXL (Fuel-Wood)	3.32 kW
2.	Ojas—M06 (Fuel-Pellets)	5.43 kW

Source: MNRE. Available at http://www.mnre.gov.in/schemes/decentralized-systems/national-biomass-cookstoves-initiative/. Accessed on 23 November 2014.

and then have several bigger joint exercises. There is a Global Alliance of cookstoves. They have a plan to assist a roll out of 100 million stoves globally. Perhaps they could help in this task, but this should be government funded. This could lead to test up to a million stoves in the next 3–4 years in phases. Only then should a bigger programme be rolled out. But the first requirement would be the commitment of the government to such a plan, which should be a 10-year plan hoping with different phases. This must plan to cover 1 million stoves by the fifth year. Dedicated funding should be provided and assured after proper study and evaluation of the needs. We have mentioned the need for a separate agency for biogas. The same authority can also look after the cookstove programme. In fact, we want to broaden the mandate and include rural electrification through renewable energy sources also for it. This authority should also be able to rein in support from the Ministries of Health and Rural Development, the two bigger stakeholders, who somehow are absent from the scene. Needless to say, the authority must be constituted immediately.[27]

In the meanwhile, in June 2014, MNRE has rolled out a new initiative for the 12th Plan period 2012–17. The target is to put in place 2.4 million household biomass cookstoves by 2017. Of these, it is expected that half the number would be supplied by the manufacturers and half would be earthen stoves installed at the household level by trained masons. The target for 2012–14, the two years which have passed was 0.1 million stoves. It is proposed to do 1.5 million during the period 2014–16. The manufacturers are supposed to give two years warranty. There probably is no stove yet available where such warranties can be given. The subsidies available for household stoves are ₹800 for forced draft, ₹400 for natural draft and earthen cookstoves with metal combustion chambers. The household stoves are supposed to be provided through women's self-help groups wherever they exist. In other places, state agencies would provide. It would have been appropriate to test different stoves in field conditions at first, which would eliminate performance risks in a larger programme. We will have to wait and see the outcome of this initiative.

Notes

1. On average, each premature death is associated with close to 20 life-years lost. Estimates for the health impact of outdoor air pollution suggest close to 3 million premature deaths per year and some 23 million DALYs. The health impacts of indoor and outdoor air pollution are not additive. DALYs: 'Disability-adjusted Life Years are units for measuring the global burden of disease and the effectiveness of health interventions and changes in living conditions. DALYs are calculated as the present value of future years of disability-free life that are lost as a result of premature death or disability occurring in a particular year. DALY is a summary measure of population health and includes two components, years of life lost due to premature mortality and years lost due to disability'. (*Source*: International Institute for Applied Systems Analysis, 2012).
2. Lancet Global Burden of Disease Study, 2010, *Lancet*, 380, 2224–60, Elsevier, December 2012. Available at http://www.thelancet.com/themed/glbal-burden-of-disease. Accessed on 23 November 2014.

3. Deepak Gupta had visited areas towards Sawai Madhopur in 2012 in Rajasthan and found that it was getting increasingly difficult to collect fuel wood and so people were using dung. There is likely to be a similar problem in the Deccan Plateau and many such areas.
4. Locating Gender in the Twelfth Five Year Plan Approach: Issues Emerging from a Gendered Analysis; prepared by the Working Group of Feminist Economists in the context of the Twelfth Plan Approach and shared with the Planning Commission 2012. Available at http://www.unwomensouthasia.org/assets/WGFE-layout.pdf. Accessed on 23 November 2014.
5. World Health Organization. Available at http://www.who.int/nmh/events/ncd_action_plan/en/index.html. Accessed on 23 November 2014.
6. World Health Organization. Available at http://www.searo.who.int/mediacentre/events/governance/rc/66/r6.pdf. Accessed on 23 November 2014.
7. As referred in Mission Document: National Mission on Decentralized Biomass Energy for Villages and Industries (January 2006).
8. Wasteland Atlas of India, Department of Land Resources, Ministry of Rural Development.
9. Indian Institute of Science, Bangalore, 2005.
10. The final report of the Expert Group on low carbon strategies for inclusive growth. Planning Commission, New Delhi, April 2014.
11. Ministry of New and Renewable Energy, Annual Report 2001–02.
12. Deepak Gupta was a District Magistrate at that time in Bihar. NPIC had become part of the 20 point programme of important items which were reviewed regularly and reports sent religiously to higher authorities, an important part of any target driven programme! Instinctively, he was very much in favour of this programme, though it was clearly of low priority generally, and gave some time and attention to it, even making field and house visits. Regrettably, however, the Collectors, etc., were never sensitised to the needs and objectives of the programme. So awareness was lacking, not only among the beneficiaries, but also amongst the implementers. It was found that people were responding favourably. High levels of subsidy were never discussed. There was concern even then about the limited durability of the stoves, and whether they were really meeting the needs of the people or being a little bit thrust upon them—issues which appear very relevant today. Looking back, there is a confidence that if another national programme were to be designed well and taken up as a separate mission, people in the field would be able to deliver it successfully in a few years' time.
13. Based on Zhang et al. (2006) report.
14. In 2010, MNRE fully funded a pilot programme for demonstration of both fixed and portable larger stoves for Anganwadis, Schools (for cooking mid-day meals) and Tribal Hostels. This proved to be very successful. The report is available in the public domain (http://www.inspirenetwork.org/mnre/Publications.htm, accessed on 23 November 2014). A larger programme to install 5,000 community cookstoves for demonstration in Mid Day Meal Scheme/Tribal Hostels/Government Institutions was approved for funding in 2011 from the National Clean Energy Fund, but was unfortunately not implemented. The Departments of the concerned Institutions in the Central Government have been repeatedly requested to include improved cookstoves at such places where LPG is not used in order to reduce consumption of wood, often purchased, and reduce smoke. However, this has not happened. Since there are thousands of such institutions in the country, and kitchens are being built in many states, a separate programme would be very much in the national and public interest.

15. Deepak Gupta joined the Ministry of New and Renewable Energy as Secretary in mid-2008. He started looking afresh at the issue of energy access in both respects. Even as the biogas programme was sought to be scaled up with the existing though limited resources, renewed attention was also given to improved stoves. But the support and encouragement given from the PMO was essential. This will still be required for major initiatives.
16. Before Deepak Gupta joined the MNRE as Secretary, he had also been fed with comments on non-functionality of such plants leading to wastage of efforts and resources. Soon after joining the Ministry he went to many rural areas to see these plants and found the functional and not-working plants. Actually, the experience suggested this programme needed to be pursued with vigour, which was done and targets increased.
17. Based on discussions with MNRE.
18. PIB, 27 June 2009. Available at http://pib.nic.in/newsite/erelease.aspx?relid=49446. Accessed on 31 August 2012.
19. Ministry of Petroleum and Natural Gas order no. P-20020/22/2009-Mkt dated 6 August 2009. Available at http://petroleum.nic.in/RGGLV.pdf. Accessed on 23 November 2014.
20. Ministry of Petroleum and Natural Gas, Annual Report 2009–10 (p. 10). Available at http://petroleum.nic.in/Annual_Report/AR09-10.pdf. Accessed on 23 November 2014.
21. Rajya Sabha Unstarred Question no. 2752 replied on 18 December 2012.
22. Ministry of Petroleum and Natural Gas. Available at http://petroleum.nic.in/report-price.pdf. Accessed on 23 November 2014. It is a different matter that the subsidy continues to be enjoyed by the urban middle class.
23. Presentation on 12th Five Year Plan. Available at http://www.slideshare.net/PlanComIndia/energy-sector-in-the-12th-plan. Accessed on 23 November 2014. In the 12th Five Year Plan document, there is no mention of LPG connections to be provided. However, the report states that in the Plan terminal year 2016–17 (http://planningcommission.gov.in/hackathon/Energy.pdf, accessed on 23 November 2014), LPG consumption will be around 22 million metric tonnes only. This implies that if on a normative basis, one household consumes one 14.2 kg LPG cylinder per month, then 128 million connections could be given. The estimates of 102.5 million might be an estimate that also takes into consideration the alternative LPG uses.
24. *Kashmir News Specials*: Ladakh's giant leap in renewable energy. Available at http://www.kashmirnewz.com/f000121.html. Accessed on 23 November 2014.
25. During visit of Deepak Gupta to interior areas of Ladakh in 2012 near and beyond the Pangong lake he saw many houses with such cookers.
26. Dr Kirk Smith of UC Berkeley, USA and Dr Rajendra Prasad, IIT Delhi.
27. A proposal had been initiated in 2010 for a Section 25A Company, but it has not progressed.

8
Subsidies and Funding

This chapter explores issues related to fossil fuel subsidies and their possible distortions in the context of rural energy requirements. It also explores funding possibilities for large upscaled programmes.

Cooking Energy: Inter-fuel Substitution in Rural Areas

This brings us to the larger question of the problems of inter-fuel substitution for cooking in rural areas. This was the main question addressed in a World Bank study over a decade ago.[1] The foregoing discussion is based on its observations and findings, which have perhaps become even more relevant today. The basic conclusion is that the effectiveness of fiscal instruments, such as changing relative fuel prices or increasing income relative to fuel prices, in promoting a switch from traditional biomass to petroleum cooking fuels in rural areas would have serious limitations. Analysis of household fuel choice in India, examination of alternative policies to the current subsidy schemes, and a review of international experience suggest that it would be difficult, if not impossible, to design an effective subsidy scheme for kerosene or LPG. Worldwide experience shows that it is extremely difficult to use subsidies to induce the poor to switch to kerosene or LPG for cooking from biomass.

Difficulties of LPG

There is an analysis of the Deepam scheme, which was piloted in Andhra Pradesh in July 1999. This had LPG subsidy and also waived cylinder deposit fee for mostly women beneficiaries of SHGs. Assessment revealed that most households found it difficult to manage the cash payments for cylinder refills even with the large subsidy prevailing at the time, resulting in incidental use of LPG for making tea or preparing meals for unexpected guests.[2] In general, the operational costs, being much more than start-up costs, limit the effectiveness of subsidising the start-up costs. While the scheme facilitated the uptake of LPG, it failed to encourage the substantial and sustainable use of LPG by its intended primary beneficiaries, the rural poor. The review found that biomass remained the main cooking fuel for most beneficiaries, especially for the cash-strapped rural households that could not easily afford the relatively high cost of LPG refills.

What distinguishes LPG from other fuels is cylinder management. This has two aspects—initial costs and problem of refills. Because LPG has to be stored under pressure, metal cylinders are required. The initial deposit fee for a cylinder and the purchase cost of an LPG stove may be in excess of ₹1,500. The combination of the start-up cost and the cash outlay at each refill (which typically cannot be broken up into smaller instalments) presents a serious barrier to the uptake and regular use of LPG by low-income households. The second problem is assuring the regular supply of refill cylinders. For small and remote markets, refills may be delivered once a week or 2 weeks—this could mean that a household may have to go without gas for 2 weeks or have 2 cylinders, which increases the cost. Infrequent delivery serves as disincentive against entirely switching to LPG. Smaller cylinders have not solved the problem. Rural markets, low population density, poor road infrastructure, low LPG uptake and low consumption among those who sign up make it difficult to establish a commercially viable distribution framework.

The study calculated the cost of cooking at that time. At a subsidised retail price of ₹469 per cylinder, the cost would come to ₹235 per month if a cylinder is used for 2 months. Cost from kerosene at subsidised price of ₹16.54 would be about ₹172. The cost of purchased wood at prices in 2002 would be ₹110. This is theoretical at best because this kind of subsidy of ₹450 per cylinder would be completely unsustainable for the government, which is struggling to reduce this in urban areas. The rural household also cannot buy 20 litres of kerosene from the PDS as this quantity is simply not available nor can be. The real market price for both kerosene and LPG would also make both completely unaffordable for most rural households. The trend would be for cost of LPG and kerosene to become higher, even with subsidies. And for most people biomass is free. Besides, rural households would also be affected by volatility in oil prices and consequent frequent LPG prices.

Kerosene

As far as kerosene is concerned, it has not been possible to identify a viable mechanism to better target and deliver the kerosene subsidy anywhere in the world. The subsidy is inherently prone to significant leakage, as has been found consistently in countries with such subsidies, because kerosene is the perfect substitute for automotive diesel. Subsidies need to be sizeable to induce the poor to take up kerosene, but a large subsidy would increase the leakage and be fiscally unsustainable. LPG is strongly favoured by the rich as a cooking fuel. Any subsidy for these fuels, regardless of its design, therefore, is subject to significant leakage, mis-targeting or both. We will talk of kerosene in more detail later.

Cash Transfers

Cash transfers to the poor to compensate for the subsidy phase-down or elimination is a sensible policy because it gives freedom of choice to the recipients. While it should help in doing away with bogus beneficiaries or diversion, and should on that account be further expanded, especially in the towns of the states with higher allocation and much better rural electricity supply, it does not seem suitable for promoting a shift in cooking fuel use towards more expensive clean fuels, particularly in rural areas. Modelling of the NSS data, consistent with international evidence, indicates that rural households conversely may use more wood if a modest amount of cash is given to them. Coupon scheme also failed as shown by Nepal/Peru experiences. In fact, it is also seen that as rural households become richer, their total energy consumption rises, resulting in an increase rather than a decline in wood consumption. This trend is also consistent with observations in other countries.

Income Distribution

There is an interesting study, which concludes that although the shares of electricity and LPG increase in income deciles 9 and 10 in rural areas, biomass remains the predominant form of household energy use, even for the highest income groups (Figure 8.1). In general, those who used wood were poorer than those who used LPG. However, there are exceptions that show that in some cases other factors apply rather than only the cost.

Behavioural Change

There are some studies regarding household behaviour with respect to fuel use change and what could motivate households to switch cooking technologies but there is little systematic evidence regarding which factors will determine this. Zhang has explicitly modelled household behaviour regarding the energy technology choices based on their attributes, including cooking costs (including stove cost and fuel cost), convenience and cleanliness (Zhang, 2010). The study found that the marginal utility of income decreases as income increases and that this effect carries over into the cooking technology choice. Thus, households are less sensitive to cooking cost as income increases. The study simulated that rural households barely change their energy technology choices if the LPG stove cost is reduced by 50 per cent. But if income is doubled, 14–24 per cent of rural households switch their primary stove from a traditional stove to a clean stove depending on their residence. This result is consistent with the conclusions drawn by several other studies that fuel switching on a large scale will not occur in rural areas unless rural economies become substantially more developed.[3] Households are more likely to choose energy technologies with shorter cooking times in areas with higher wage rates for unskilled women. With respect to cleanliness, the study showed that residents of households that know IAP is harmful to health are more likely to choose energy technologies with lower pollution levels.[4]

Figure 8.1
(a) Total rural energy consumption showing source share, by income decile and (b) End-use energy consumption showing source share, by income decile

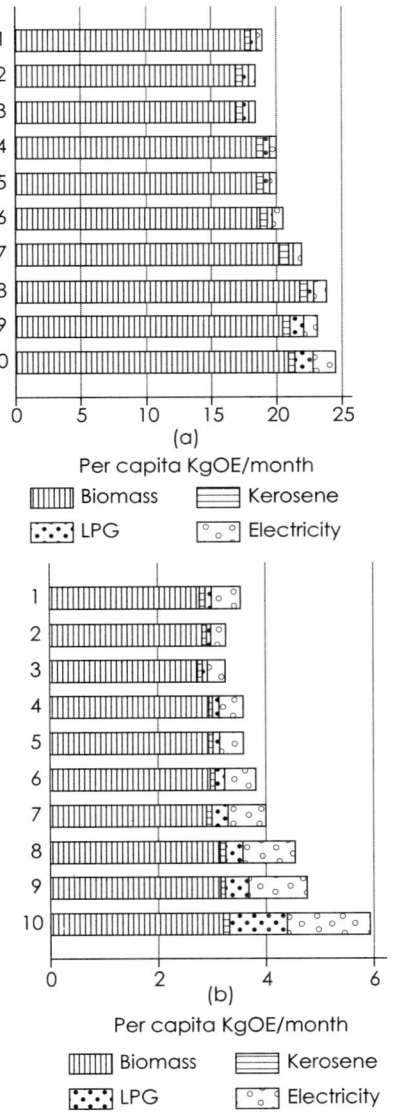

Source: Khandker et al. (2010).

All this highlights the fact that switching to cleaner fuels is not a budget priority for many households, and that, in combination with other mitigation measures, raising awareness about the health benefits in a language that poor people can understand and appreciate could be one of the most effective interventions to facilitate fuel switching. Households may realise the immediate benefits of having less smoke, but unlike experts, who are beginning to realise the long-term health consequences of indoor air pollution, households generally do not make these connections, just as most men don't change their behaviour regarding smoking. Households need to perceive, and to be convinced, about these direct and indirect benefits. It has been suggested, therefore, that promoting behavioural change may be the most promising option for those who cannot afford cleaner fuels. Accelerating the viable expansion of rural electrification is of special importance because in addition to reducing the need for kerosene for lighting it has in a number of countries been found to be strongly correlated with the uptake of clean fuels for cooking. The use of dung leads to the highest exposure level, which is nearly 50 per cent higher for the cook than that due to wood use. This suggests that fuel switching within biomass from dung to wood alone may bring about health benefits. This aspect also requires attention in any strategy. Finally, improving the overall rural economies, particularly for women's employment opportunities, and overall economic development will play an important role. Thus, energy access, for both electricity access and cooking, is bi-directionally related to general rural development, being both cause and effect.

Conclusion

Any technological alternatives to free or cheap traditional biomass will be widely adopted only if the incremental costs are affordable and outweighed by tangible non-monetary benefits valued by the user. Whether households use free or bought firewood is an important question. Where time is unconstrained (i.e., valued at zero in monetary terms) and there is firewood to be grown or collected, it is difficult for commercial fuels to compete with firewood. The same applies to dung, which is freely available to those households with cattle. In short, supply conditions in rural areas favour the use of biomass for cooking because of its low labour costs and the ready availability of free biomass. Those who rely primarily on purchased wood are the most likely candidates for fuel switching. However, the higher cost of cooking with LPG (₹176/ month) as compared to purchased wood (₹137) is the major reason for not switching. They may also be the best candidates for an improved stove, which will further reduce the cost of wood.

As we have seen, almost 86 per cent of all rural households were using solid bio fuels in 1999–2000, while the use of LPG and kerosene as the primary cooking fuel was essentially non-existent. It is no surprise that this has not changed much even now. This applied across all income groups except the richest 10 per cent. It is also unlikely to change much and biomass is likely to remain for a long period the primary source of energy for cooking needs in rural areas. In fact, even for those poorer urban or peri-urban households that cannot afford LPG or kerosene but purchase wood for cash, improved cleaner and more efficient biomass stoves and fuels (such as biomass briquette technologies) may be a cheaper attractive option. Once we recognise these realities, the problem of cooking energy access can be better addressed by making biomass central to any strategy for determining rural cooking energy solutions. Biomass is renewable, climate change friendly and if consumed with care, its ill-effects such as excessive smoke and release of particulate matter in the atmosphere, can be nullified. Some feel that free biomass acts as a barrier to the penetration of efficient and clean energy resources and technologies. But perhaps the opposite is true, since we simply cannot provide or afford to provide LPG to all people.

In addition to the health and economic empowerment benefits associated with their use, clean cookstoves and fuels can reduce a large share of emissions from cooking with biomass. Cookstove projects' climate benefits are not limited to carbon dioxide; they can also significantly reduce emissions of black carbon, carbon monoxide and total non-methane hydrocarbons. However, these benefits are not yet quantified by the methodologies, nor can credits be earned for them. These reductions also bring other benefits, such as reduced indoor and outdoor pollution, less pressure on forests, and economic and time savings due to the reduced need to search for or purchase costly fuels. Since they have short life spans—a few days for black carbon, a decade for methane— reducing these gases would bring about a more rapid climate response than reductions in CO_2 alone (Lee et al., 2013).

Funding and Subsidies

World Bank View

This brings us to the important discussion on funding and subsidies, both for improved cookstoves and biogas. The World Bank view is that there should generally be no direct subsidies for stoves as they do not help in developing a self-sustaining commercial market.[5] They say that general experience has shown that across the board, consumer fuel subsidies are not a good way of helping the poor. Affluent households tend to benefit the most from prevailing fuel subsidies, given that in most cases, energy consumption increases in parallel with income. For governments, these subsidies result in heavy fiscal deficits diverting direct public expenditures away from productive and social sectors. Alternative options are usually designed in the form of social protection programmes. The challenge remains in successfully implementing these options to effectively reach the poor. They point to the huge subsidies in NPIC, which was one of the reasons it could not be sustained and led to many other problems. Subsidies diminish acceptance beyond the specific programme. They argue that subsidies should be for other activities which help develop the market, such as in development of designs, products, market surveys, test projects, technical and testing support, awareness drives, evaluation and monitoring.

Justification for Subsidy for Cookstoves

All these arguments are well taken, and there is full agreement with all the suggestions regarding indirect subsidies. In India, however, the above is an argument against the huge subsidies for LPG and kerosene, rather than the tiny amount in comparison which will be required for improved cookstoves, which would not disturb, but rather help reduce, the fiscal deficit. The fact is that LPG is heavily subsidised and is being consumed also by higher income groups as well as most of the urban areas and the richer people in the rural areas. Petroleum Planning and Analysis Cell analysis reveal during the period from April to June 2013 under-recovery from domestic LPG was ₹9,285 crore.[6] Thus total subsidy could be in the region of ₹40,000 crore annually. These subsidies are also on a recurring basis. Therefore, it is simply not possible to argue in the

Indian situation that improved stoves should not be subsidised. It would be highly equitable to provide a reasonable subsidy to all the poor rural households who will mainly be using these stoves while simultaneously reducing the subsidy on cooking gas, and even kerosene. Savings accruing in subsidies would be far more than the extra subsidy expenditure on cookstoves. An International Monetary Fund (IMF) study suggests that energy subsidies reinforce inequality. The largest beneficiaries are upper-income households, who are the largest consumers of energy. On average, the richest 20 per cent of the population in developing economies capture six times more in fuel subsidies than the poorest 20 per cent (Granado et al., 2010). The subsidies will also be efficient as admitted by the Bank Study because they will help disseminate stoves, which have significant benefits as compared to the costs. The only question then is what should be the quantum and how they are administered so that they promote and not hinder market development.

A good durable stove with high efficiency and a life of 5 years may cost about ₹3,000 (US$50) at this stage, which could come down to ₹2,000 and possibly less with large scale-up. It would be perfectly reasonable to provide a 50 per cent subsidy for this. This could be done in two ways. The first is a voucher that will allow the beneficiary to buy a certified stove from the marketplace with a rebate. This will help develop the market. The alternative is to convert it into an interest subsidy for bank/micro/SHG finance, which will allow the stove cost to be paid for in instalments. This will also allow for stoves to be purchased through loans from the market. In either case, subsidies will strengthen development of the market. The subsidy can be reduced, or even eliminated, in future replacement cycles.

Subsidy for Biogas

Criticism of Subsidy

Let us look at the issue of subsidies for biogas. There is criticism that this is a wasteful subsidy, implying thereby that it is going down the drain as the plants are not working. We think this is both misinformed and biased. Urban middle classes and the rich continue to get huge, and recurring, subsidies for cooking gas, which is not only acceptable but where people take to the streets when subsidies are sought to be reduced. How then can we criticise a one-time subsidy for a biogas plant, or a subsidy that might be repeated after 15 years? As regards functionality, we have addressed this issue earlier. This criticism is perhaps made by those who have never visited the field.

Comparison with LPG Subsidy

It would be useful to do a comparison of these subsidies. The estimated biogas generation from the biogas plants installed in the 11th Plan period comes to about 15.20 lakh cubic metres per day equivalent to a daily saving of about 13.5 million LPG cylinders per year, besides producing about 5,500 million kilograms of nutrient-enriched organic manure per year, replacing 58.7 million kilograms of urea per annum. Based on December 2013 selling price of a domestic LPG cylinder,[7] with restriction of nine cylinders a year, would show that the current proposed average subsidy of ₹12,000 per plant is equivalent to actual LPG subsidy of little over 2 years. In addition, there would be savings in subsidy for the urea saved. An economic analysis would note the positive externality benefits of quantities of cooking gas and urea actually saved. Further, the independent evaluation observed that, on the average, earnings from sale of such manure from a biogas plant was about ₹30,500 per year. It thus becomes an income-generating or cost-saving activity for the beneficiaries.

Savings

It is clear then that there are huge savings in subsidies over the years if biogas plants were to be developed. Admittedly, biogas is not for the poor because people must have sufficient cattle and they should be able to have them stall fed. It is these people who would also be able to shift to LPG should it be easily available and with subsidy. For them, biogas becomes a much better alternative. It is also surprising that other benefits of biogas are hardly talked about, or perhaps they are not well understood. If we look at carbon emissions, one family type biogas plant that on normative basis replaces one LPG cylinder each month abates around 3.5 tonnes of emission annually. So, when we are looking at all avenues for reducing emissions, this is a good alternative. In many cases, it may lead to less dung being used for cooking, which is now a fairly common practice in Uttar Pradesh and Bihar. Since dung as such has the worst adverse health effects and is not so good for cooking, conversion of dung for cooking to biogas has commensurate benefits, which should also be monetised. Finally, good fertiliser output results from the process, which can be used by the beneficiary for his kitchen garden or the fields or sold. This is a win–win situation, but hardly recognised as such.

Financing of Subsidy

Requirement of Funds

Now let us see what could be the costs of such a subsidy. Let us say we are talking of 100 million stoves over a 10-year period. This is the original market not the replacement one. If each stove has a subsidy of about ₹1,000 in rupee terms, it would mean a subsidy of ₹10,000 crore over a 10-year period. The development and public awareness costs may be another 1,000 crore over this period. If successful, with this kind of support, a huge market will develop because it would mean 10 million first-time stoves annually on the average, and later after 5 years of the start, there will be an equivalent replacement market each year. With the development of newer models and the briquette market we may even touch more numbers by tapping into the urban peripheries, if and when LPG becomes difficult to supply or very expensive. It is hoped that the experience with stoves would be such that people would like to buy the next stove without any subsidy, or perhaps the subsidy can be reduced by half for the second stove. We are giving approximations because such a huge market may reduce manufacturing costs. Imagine the employment created in the whole manufacturing and distribution system and the economic opportunities for those involved in that system. In the case of biogas plants, if we were to install 1.5 lakh plants a year, it would mean an annual subsidy of ₹180 crore per year or only ₹1,800 crore over 10 years.

Meeting the Requirements

A question may be asked as to where this money would come from. In a country where we are spending ₹40,000 and ₹29,000 crore, respectively, on annual subsidies for LPG and kerosene, most of which can be done away with, and a large portion of kerosene subsidy currently being used for adulteration, this amount of ₹1,100 crore per annum for people at the bottom of the pyramid and ₹180 crore per annum for slightly better off rural folk should not be a problem at all. About 15–20 per cent of the current annual coal cess proceeds (which will actually substantially go up in the next few years with an increase in the use of coal for power), now going into a dedicated National Clean Energy Fund, will be able to fund this entire enterprise. The existing pattern of distribution of subsidies in some of the developing countries, including India shows there is enough slack to provide for this. A simultaneous cut in fossil fuel subsidies would also more than provide for this (IEA, 2011).

The Carbon Market

Potential Carbon Revenues

There is another alternative, which unfortunately the international system has failed to provide. The various World Bank studies, and others, have pointed out the possibility of getting private and public carbon funds. But, apart from a few pilots here and there, there is little of substance. There is one example in India. Agriculture Development and Training Society (ADATS) and Bagepalli-based NGO along with Coolie Sangha, an independent community-based organisation, got a clean development mechanism (CDM) project registered in August 2009 to set up 18,000 biogas plants of 2 m^3 capacity each for rural households in this district.[8] There are some successful programmes in some African countries but at relatively low scales. As IEA noted, as of June 2011, there were only 15 CDM projects or 0.2 per cent of the total related to improved energy access for households.

But let us look at the economics and the procedure. An improved stove should provide about 1.5 CER annually. With a life of 5 years this becomes 7.5 CER. At a reasonable price of Euro 8 per CER, it has potential to generate carbon revenue of ₹4,800 (@ ₹80 per Euro) over five years. This money is sufficient to meet the subsidy to the beneficiary and all other costs including the setting up of a proper distribution and monitoring system and carry out public awareness programmes and evaluation etc. This could really become a self-financing programme. A transfer of resources from the National Clean Energy Fund to the proposed Authority would be the easiest way in India to allow for the front loading which would only have to be done for the first 50 million stoves over the first 5 years. Thereafter, revenues would make it self-sustaining.

In the case of biogas, one biogas plant of 2 m^3 could earn about 3.5 CERs in a year. In 15 years, it could earn 52.5 CERs or 525 Euros @ 8 Euros per CER. This would mean ₹33,600. Therefore, if a biogas plant is operated and maintained well, we could reduce the current subsidy by half and meet all the costs of a biogas programme with much reduced beneficiary contribution too. Here also there will be the issue of front loading of costs and flow of annual benefits.

In earlier years, the CDM procedures were just too difficult and the transaction costs too high. Later, with the introduction of the programmatic CDM, these issues became more manageable to some degree, even though more reform was required. This was why MNRE decided to register a cookstove programmatic CDM under the Framework Convention on Climate Change (FCCC). Ministry was unable to provide funds. GIZ agreed to get this done. The CDM Executive Board registered the project under the title 'National Programme for Improved Cookstoves in India' on 28 December 2012.[9] A couple of private enterprises have also projects registered, but their charges for inclusion of others appeared to be too high.

Carbon Market Alternatives

Unfortunately, the carbon market has since crashed, making this alternative almost redundant. It is also unlikely to recover soon. The uncertainty on the future of CDM like mechanisms suggests little expectation of financing from this source. It is also possible that countries such as India may not be allowed benefits from it, even though such projects are for people at the bottom of the pyramid. However, there is still a voluntary market and a gold standard. There must be an international market place that provides a platform for sellers and buyers to get together. If voluntary carbon market is limited it will be able to account for only a limited amount of emissions. This being so, emissions saved through use of improved stoves would probably constitute the best possible programme to fund in this manner. So we raise the following questions/make the following suggestions:

- Can there be a dedicated CDM for cooking energy emissions saved and available at a minimum normative CER value of Euro 10 per CER?
- If not, can the World Bank host a platform and locate dedicated funds under the voluntary carbon market?
- Can the proposed International Climate Fund help in nations creating dedicated revolving funds by making equal contributions to meet upfront costs? If so required, such funds can charge an interest of 1–2 per cent and the principal to be returned after 10–15 years.[10]

Were any of these to happen, the financing costs for providing substantive cooking energy solutions to rural areas would be easily met.

International Funding

IEA has analysed in detail the finances required to make a substantial difference by 2030 (IEA, 2011). It suggests that US$21 billion will be required from 2010 to 2030 to provide 860 million people with clean cooking facilities. This is equivalent to an average annual investment of US$1 billion to provide facilities to about 41 million people per year. It further talks of possible funding through the carbon market, multilateral and bilateral agencies, private sector, etc. Cooking energy is not an area where the private sector will easily come in because there is no income to generate. This could change if there is a robust voluntary carbon market. It is clear that cooking energy will require funds from the public sector and/or international funding. Ideally, there should be a dedicated window on the proposed Climate Fund. In fact, this could be separated from the larger fund. Hopefully, when the international system talks of the unacceptable tragedy of 2.8 billion people using biomass in traditional stoves for cooking, it will also be actually able to quickly set up such a small dedicated fund! We believe that something very different must be done, and urgently.[11] We made this suggestion in Chapter 6 also for electricity access. Both actually go together.

Domestic Funding

In view of the evident difficulties of funds coming from these sources, it appears that India would have to find its own solutions and fund them from its own budget. This is also what appears to be the conclusion from a reading of the IEA analysis. In India, for cooking, they propose a huge growth of biogas (60 per cent), with improved cookstoves and more gas stoves being 20 per cent each—all funding coming from beneficiaries with some public sector support. Such a large increase in biogas is unlikely, and the numbers will probably be made up by increasing the percentage of improved stoves.[12] And we have seen, funding for cookstoves can easily come from simply slight restructuring of existing subsidies, and the rest would come from the people themselves, perhaps helped by bank finance. It may be worthwhile to study the Sri Lankan model where reportedly 6 million improved stoves have been installed.

The Kerosene Conundrum

Census 2011 revealed that around 78 million, or 31 per cent, of Indian households use kerosene to meet their lighting requirements. Of these, 73 million are rural households. Kerosene-based lighting devices include kerosene wick lamps, hurricane lanterns, kerosene petromax and non-pressure mantle lamps. Households in non-electrified areas or non-connected households use kerosene for lighting, but electrified households in rural areas also use kerosene as a backup fuel because of erratic and poor electricity supply. We have noted earlier the very small percentages of households using kerosene for cooking both in rural and urban areas. Essentially, therefore, the role of kerosene is for lighting in rural areas.

The Ministry of Petroleum and Natural Gas allocates kerosene quota to the states. It supplies kerosene to the State Departments of Food and Civil Supplies through the oil companies as per the allocation, which in turn distributes it to the PDS. On the other hand, electricity distribution is under the purview of the Ministry of Power and, in turn, state utilities. Thus, there is no coordination regarding level of electrification and availability of electricity with allocation of kerosene quota. Lighting needs, and in which areas, do not figure in the criteria for allocation. Of course, it would be interesting to see how allocations would change if indeed such coordination was there.

The allocation of subsidised kerosene by the Central Government varies from state to state and has been based on historical patterns rather than on demand or on consideration of relative poverty levels. Quantities allocated to states evolved based on actual take off by states over the years. Different states use different criteria for PDS supplies, so the per capita basis varies widely across states. Allocations have generally been much higher for richer states reflecting total off-take of kerosene rather than that by the poor. Since 2005 kerosene consumption in the country by and large remained at the same level but there has been some welcome reduction in allocations in the last few years, perhaps because of linkage with increase in LPG connections or otherwise (Table 8.1). The total reductions have been about 8 per cent in the last two years each. There have been substantial reductions in the case of Maharashtra and Gujarat. It seems that the criteria continue to be cooking needs for kerosene. There is an urgent need to link kerosene allocation with lighting needs and to peg it to figures of electrified villages, or at least with electrified households with some criterion for cooking. There appears to be thinking in the new government along these lines. If this were to be done, then there could be a drastic reduction in allocations and a judicious and justifiable redistribution of quota amongst states. There could be an annual reduction of 15 per cent concentrated in a few states while the issue price could be increased more slowly at 10–15 per cent change every 6 months. As solarisation of villages/households takes place, allocations could be reduced in tandem.

The total PDS kerosene consumption in 2010–11 was around 11.34 billion litres.[13] On the basis of provisional population figures of Census 2011, the per capita allocation of PDS kerosene at the national level works out to 8.6 litres per annum. The per capita allocation of PDS kerosene for non-LPG covered population is 19 litres per annum.[14] The average annual per household kerosene consumption on one kerosene lamp that is lit for 4 hours a day is around 60 litres per year, or on an average of 5 litres per month.[15] This would normally be lower for poorer people. But even on this normative basis, the annual kerosene consumption from 78 million households would be 4,680 million litres. This is

Table 8.1
State-wise kerosene allocation (in MT)

S. No.	State	2005–06	2006–07	2007–08	2008–09	2009–10	2010–11	2011–12	2012–13
1.	Andhra Pradesh	518,014	524,958	517,936	517,158	527,181	463,532	413,023	362,643
2.	Andaman & Nikobar	5,816	6,407	6,797	5,816	5,658	5,639	5,640	5,631
3.	Arunachal Pradesh	9,257	9,257	9,257	9,257	9,169	9,131	9,048	8,993
4.	Assam	258,396	258,007	263,007	258,007	257,875	258,044	257,326	255,371
5.	Bihar	647,430	647,430	662,994	657,430	643,743	641,663	638,295	635,963
6.	Chandigarh	13,067	13,067	13,067	9,999	7,180	7,133	5,705	3,082
7.	Chhattisgarh	146,938	146,938	146,938	146,938	145,812	145,464	145,194	144,131
8.	Dadra & Nagar Haveli	2,782	2,782	2,782	2,782	2,785	2,362	1,933	1,774
9.	Daman & Diu	2,118	2,118	2,118	2,118	2,072	1,811	1,569	710
10.	Delhi	168,484	168,484	168,484	160,935	135,226	108,064	47,758	41,949
11.	Goa	19,212	19,212	19,212	19,212	19,208	17,645	15,388	4,249
12.	Gujarat	744,381	747,650	743,759	743,759	742,813	716,193	524,114	524,190
13.	Haryana	145,619	145,619	145,619	145,619	144,820	134,385	122,364	73,989
14	Himachal Pradesh	50,537	50,537	50,537	49,409	45,463	31,322	25,266	19,564

(Table 8.1 Continued)

(Table 8.1 Continued)

S. No.	State	2005–06	2006–07	2007–08	2008–09	2009–10	2010–11	2011–12	2012–13
15.	J&K	79,078	76,433	76,433	76,200	75,554	74,208	74,220	73,695
16.	Jharkhand	211,175	211,175	211,175	211,175	210,950	210,723	210,303	210,107
17.	Karnataka	461,867	462,412	461,867	461,478	465,199	437,868	419,821	406,917
18.	Kerala	216,308	216,308	216,308	216,308	216,296	175,125	153,383	97,429
19.	Lakshadweep	795	795	795	795	795	794	794	1,018
20.	Madhya Pradesh	489,017	489,231	488,609	488,609	487,823	487,349	487,414	487,144
21.	Maharashtra	1,284,658	1,280,768	1,276,876	1,276,876	1,276,502	1,216,929	979,482	735,969
22.	Manipur	19,907	19,907	19,907	19,907	19,742	19,718	19,720	19,723
23.	Meghalaya	20,401	20,401	20,651	20,401	20,358	20,334	20,281	20,190
24.	Mizoram	6,217	6,217	6,217	6,217	6,180	6,162	6,097	6,098
25.	Nagaland	13,312	13,612	13,312	13,312	13,317	13,304	13,306	13,307
26.	Odisha	314,977	316,144	314,977	325,172	314,312	313,643	312,145	311,104
27.	Puducherry	12,296	12,257	12,257	12,335	12,248	12,239	8,123	3,632
28.	Punjab	237,192	237,192	237,192	237,192	234,685	222,038	212,077	80,844
29.	Rajasthan	398,913	401,247	401,248	398,913	398,404	398,059	397,926	397,634
30.	Sikkim	5,582	5,582	5,582	5,582	5,566	5,135	5,360	4,940

31.	Tamil Nadu	575,271	568,267	570,602	564,075	558,391	496,479	429,413	375,287
32.	Tripura	30,832	30,832	30,832	30,832	30,738	30,575	30,552	30,490
33.	Uttar Pradesh	1,243,826	1,246,908	1,241,772	1,242,939	1,240,705	1,239,952	1,239,288	1,246,808
34.	Uttarakhand	89,985	94,517	89,849	89,849	91,784	86,405	83,662	29,519
35.	West Bengal	752,103	756,149	754,049	752,352	754,286	751,227	750,980	750,556
	All India	9,195,763	9,208,820	9,203,017	9,178,958	9,122,844	8,760,653	8,066,968	7,384,649
	Growth (%)	−0.3%	0.1%	−0.1%	−0.3%	−0.6%	−4.0%	−7.9%	−8.5%

Source: Petroleum Planning and Analysis Cell, Ministry of Petroleum and Natural Gas.

only around 42 per cent of the kerosene distributed through PDS system. Therefore, it appears that allocation of kerosene can and should be considerably reduced. This would save substantial funds as also help in rationalising the allocation process. This saved subsidy could be used for promotion of improved cookstoves.

PDS kerosene price has traditionally been kept very low. In 2002, the PDS kerosene was available at around ₹9 per litre. This price was kept at that level for the next few years even as the subsidy levels increased exponentially from ₹3,751 crore in 2003–04 to ₹28,000 crore per year in 2008–09. It is only in the last few years when a limited attempt was made to rationalise energy subsidies that the prices were partly revised. Even then, total subsidies are increasing (Table 8.2).[16]

This huge price difference creates many 'arbitrage' or rent seeking opportunities by creating perverse incentives for diversion and adulteration. This has led to the politicisation of dealership allocations because of the opportunities for providing significant amounts of rent. As these

Table 8.2
Under-recoveries of OMCs and compensation by upstream companies and the government

Year	Kerosene Subsidy PDS (₹ thousand crore)	PDS Kerosene Selling Price (in ₹)
2003–04	3,751	9.0
2004–05	9,480	9.0
2005–06	14,384	9.0
2006–07	17,883	9.0
2007–08	19,102	9.0
2008–09	28,225	9.22
2009–10	28,000	9.32
2010–11	28,000	12.37
2011–12	28,000	14.83
2012–13	29,140	14.83
2013–14 (est)	29,519	14.96
Total[17]	235,484	

Subsidies and Funding   225

linkages get strengthened and entrenched it makes it increasingly difficult to clean up (Morris et al., 2006). The brazen attacks on activists protesting the system is enough evidence of this. The rents are so high that even the increase in costs because of poor mileage and wear and tear of the vehicles because of adulteration are not constraining factors.

There have been several studies regarding the extent of diversion. In 2002, based on NSS data, an analysis suggested diversion of 50 per cent. In the year 2003, 977.29 million litres of kerosene oil was released under PDS. NSS 59th round valid for the same year accounted for 483.60 ml. Thus, 50.5 per cent was the diversion. National Council of Applied Economic Research (NCAER) data of 2005 suggested more conservative figures that diversion from PDS was about 39 per cent of which about 18 per cent got diverted back to household consumption (Morris et al., 2006). It appears, therefore, that a diversion of half the supply is actually happening, some of which comes back to households. This means that (a) there is considerable advantage in rent seeking through diversion and (b) many rural households, or possibly those without ration cards in urban areas, are actually paying market price for this PDS kerosene. Since the differentials between the subsidised price of kerosene and diesel have only widened over the years, these anomalies may have actually become worse (Table 8.3).

The subsidisation of kerosene for the economically weaker sections of society, therefore, leads to huge outgoes from the budget; reveals the failure to stop the diversion; allows huge rents to be gained by unscrupulous elements with other overtones; facilitates adulteration of other products with kerosene and resultant costs; and even leads to the wrong production mix at refineries.

Kerosene subsidy is one of the areas to be covered under the government's plan to implement the direct cash transfer system to ensure targeted delivery. It is hoped that this will lead to substantial savings in terms of reduction in consumption of kerosene by reducing the diversion. Two years ago a pilot project was started in District Alwar in Rajasthan. This showed considerable savings, though it also created difficulties for many people because of bank account opening and other problems. There are also reports of delayed payments of subsidies. This was also extended to many other districts and made a substantial difference. But its further extension was discontinued. However, this

Table 8.3
Actual retail selling price of PDS kerosene and under-recovery at Delhi

Year	Actual Avg. RSP (₹/litre)	Under-recovery (₹/litre)
2002–03	8.98	1.69
2003–04	9.00	3.12
2004–05	9.01	7.96
2005–06	9.07	12.1
2006–07	9.09	15.17
2007–08	9.13	16.23
2008–09	9.19	24.06
2009–10	9.24	14.85
2010–11	11.70	17.39
2011–12	14.34	26.46
2012–13	14.85	31.16
16-Sep-13	14.96	36.83

Source: Petroleum Planning and Analysis Cell, Ministry of Petroleum and Natural Gas.

laudable effort, which the new government is rapidly extending in its coverage, only intends to reduce the diversion, and attendant subsidy. It is, therefore, addressing the symptom, not the cure. The real issue is to reduce the actual consumption of kerosene also and consequently the subsidy support to a poor, inefficient and polluting source of lighting when environmentally benign options are available.

Of course in actual practice, this would pan out differently but the following steps could be taken in tandem, which we are discussing in more detail elsewhere. Make a plan to provide solar lights or solar mini-grids for the entire concerned population over a medium term and provide the budget for this. This money could come from the budget in the first year with additionalities every year from subsidy savings through annual reduced allocations of kerosene by 10–15 per cent. Raising the issue price of kerosene by 10–15 per cent every 6 months would also reduce subsidy outgo while making the shift to solar lights easier because household kerosene costs would increase, as also reducing

Subsidies and Funding 227

the incentive for diversion. What better win–win situation can there be because after a few years we could have—reduction in kerosene consumption by over 2–3 billion litres; reduction in diversion, adulteration and black-marketing of kerosene by perhaps more than 80 per cent; reduction in annual kerosene subsidy by almost ₹10,000–15,000 crore; creation of a solar market of a few hundred megawatt with a huge technical solar ecosystem and employment of hundreds of local technicians in rural areas.

Notes

1. Access of the poor to clean household fuels in India. UNDP/ESMAP, New Delhi, 2003. Available at http://www.cleancookstoves.org/resources_files/access-of-the-poor-to-clean.pdf. Accessed on 23 November 2014.
2. Biogas at many places is also used similarly—personal experience in field visits to many villages in different states—those households may also be using traditional cookstoves.
3. The studies quoted on page 8 of the World Bank study report are Heltberg (2004, 2005), Zhang, Barnes, Sen (2007) and Zhang and Vanneman (2008). The study is by Ekouevi and Tuntivate (2013).
4. Ibid. Ekouevi and Tuntivate (2013, pp. 8–9).
5. This point has been made strongly in Barnes et al. (2012).
6. Petroleum Planning & Analysis Cell: 1 October 2013. Available at http://ppac.org.in/WRITEREADDATA/PS_oil_prices.pdf. Accessed on 23 November 2014.
7. *Business Today*, 10 December 2013. Available at http://businesstoday.intoday.in/story/lpg-price-hiked-by-rs-3.46-per-cylinder/1/201351.html. Accessed on 23 November 2014.
8. Available at http://cdm.unfccc.int/Projects/DB/DNV-CUK1242729511.7/view. Accessed on 23 November 2014.
9. CDM Executive Board. Available at http://cdm.unfccc.int/ProgrammeOfActivities/poa_db/18TQ93F4AOIDNGYW7C6BMPKE0RJVLU/view. Accessed on 23 November 2014.
10. There is a good example in the formation of the Global Fund for AIDS, TB and Malaria which gives grants.
11. Deepak Gupta has been making these arguments and suggestions in discussions in the international fora, but there appears little movement in a positive direction.
12. There is an error in the IEA report: Energy for All: Financing access for the poor (2011) (p. 19): showing China having installed 5 million plants, and India along with Sri Lanka and Nepal having installed 0.2 million. The functional family type biogas plants recorded in India are about 1.2 million, although over the 20–30 years more than three times were installed.
13. Report of the High Powered Committee on Financial Position of Oil Companies (2008) and Note for supplementary Lok Sabha Starred question no. 49 for 24 February 2011 asked by Shri Anurag Singh Thakur and Shri Dushyant Singh regarding pilferage/black marketing of subsidized kerosene. Available at http://pptfun.

com/Indiaoilgas/MOPNG/Adulteration_Lok_Sabha_Question_49_24feb2011.pdf. Accessed on 23 November 2014.
14. Rajya Sabha question no. 136 answered on 13 March 2012 by Shri Kanjibhai Patel on 'Reduction in kerosene quota to Gujarat'.
15. Assumption for daily usage of kerosene lantern and consumption per hour. That is, specific fuel consumption of kerosene lantern @ 0.041/hour and daily usages for 4 hours has been taken from Chaurey and Kandpal (2009).
16. Report of the Expert Group on a Viable and Sustainable System of Pricing of Petroleum Products, Government of India, New Delhi, 2 February 2010. Available at http://petroleum.nic.in/reportprice.pdf. Accessed on 23 November 2014.
17. Up to the year 2008–09, the under-recovery figures have been taken from the Report of the Expert Group on a Viable and Sustainable System of Pricing of Petroleum Products, Government of India. For the year 2009 onwards, the under recovery has been calculated on the normative loss of ₹25 per litre only, although the exact figures for the period vary and in the range of ₹26 per litre plus. And in February 2012, the under-recovery per litre was ₹28.66 (http://www.rediff.com/business/report/budget-2012-heres-what-oil-and-gas-sector-expects-from-budget/20120314.htm, accessed on 23 November 2014). This under-recovery is now much more.

9
Energy Access and Rural Development

Rural development is necessarily a priority policy area for India. Although several measures have been taken towards rural development, the approach has not been as holistic as it needs to be. In this context, we look at a brief picture of what impact energy access could have on rural transformation, other than by the already established benefits of improvement in quality of life and human development possibilities.

Rural Energy Market

There can be little doubt that India will not develop significantly, nor will there be inclusive growth or poverty removal, unless there is a rural transformation. Such a transformation means not only a rise in income through livelihood development, or increased agricultural production and better local employment, but also a better quality of life, improved health and education, and greater opportunities for improving individual conditions. There is a need for food security, improved infrastructure, better seeds and agricultural practices, and enhanced health facilities. Above all, the pace and depth of rural renaissance will depend on energy access.

The growth of many sectors of the economy is driven by rural demand. More than 70 per cent of India's population lives in villages and constitutes a big and growing market for industry due to increasing disposable incomes and awareness levels. The consumption expenditure in rural India between 2009–10 and 2011–12 stood at ₹375,000 crore as against ₹299,400 crore by urban India (*Financial Express*, 2014). About one in every two rural households now has a mobile phone. Even in India's poorest states, such as Bihar and Odisha, one in three rural households has a mobile phone. Nearly 42 per cent of rural households owned a television in 2009–10, up from 26 per cent five years earlier. Similarly, 14 per cent of rural households had a two-wheeler in 2009–10, twice the number in 2004–05, though still remaining well below the urban household penetration level of 33 per cent in 2009–10.[1] Over time, a large portion of the rural population will have opportunities to move up the consumption ladder as rural areas of relatively poor states such as Bihar and Uttar Pradesh catch up with today's income and consumption levels of more affluent states.

Rural Enterprises

According to the fourth All India Census of Small Scale Industries, 2006–07, conducted by the Ministry of Small Scale Industries in 2011, rural enterprises with 7.05 lakh working enterprises accounted for 45 per cent of the total working enterprises in the medium, small and micro-enterprises sector. Further, on an all-India basis, around 61.07 per cent of enterprises were using electricity as their main source of energy. Around 3,000 enterprises—0.19 per cent—were using renewable energy as the main source of energy, and around 23,000 enterprises—1.5 per cent—were dependent on fuel wood for meeting their energy needs.[2] It would be evident that the growth of enterprises in rural areas is constrained by the non-availability of energy and also that expenditure on fuel represents a major share of production costs. Rapidly increasing prices of commercial fuels are reducing the profit margins of these enterprises. Renewable energy, on the other hand, offers many possibilities to power micro enterprises. Biomass-based power could also power small enterprises. Gasifier technology can provide dedicated loads to up to 2 MW. In paddy-growing areas all rice mills could be fuelled by their own by-product, which is rice husk, along with some diesel.

We should recall the benefits as far as micro enterprises are concerned, which are clear in the evaluation of the RGGVY. All these benefits can also accrue if limited amounts of electricity are provided through renewable

Energy Access and Rural Development 231

energy sources. In the villages where pilot projects are operational, there are many small enterprises in existence—an *attachakki* or operation of any small machine; tyre treading; cloth weaving; various kinds of shops, etc. These are the kinds of mini enterprises that can work in the normal village environment. They can also provide a better tariff, which helps the investor or the provider of the energy. Closer to the main roads, where higher levels of service or industrial enterprises can be located, it is likely that grid electricity will be available or there can be dedicated loads through renewable energy.

Small Commercial Markets

Small commercial markets form an essential part of rural livelihoods, with people from within the villages and outside engaging in small trades. There are several categories of rural markets. They are often categorised into regulated and unregulated commercial markets on the basis of the types of players who participate and run the market. Markets (locally referred to as *mandis*) are regulated by the Agricultural Produce Marketing Committee and are primarily wholesale markets located near big towns. Larger *mandis* also exist at the district level, often situated close to major district trade centres. Unregulated commercial markets include *haats, peta, angadi, hatwari, shandies, chindies,* etc. They can be further classified into periodic (weekly) and permanent markets. Most of the unregulated commercial markets in the country are not properly equipped, often lacking essential facilities such as electricity, platforms for setting up shops and drinking water. In a survey conducted in one such village market there were about 50 permanent shops and the average power consumption was around 100 W per shop.[3] Electricity is drawn either from the grid or from a diesel generator (DG), or both, and it is used for illuminative purposes (40–60 W light bulbs). In smaller markets, traders often use battery-operated rechargeable lamps. Local entrepreneurs who own small-capacity DGs provide recharging facilities for them. The average electricity requirement of markets is about 5 kW,[4] and there are around 70,000 rural markets. Electricity can cost anywhere between ₹20 and ₹30 per day. Broad estimates indicate that a 5 kW renewable energy system could meet the energy needs of a cluster with around 50–60 shops, at a much lower cost.

Telecommunications

Telecommunications services have been recognised as an important tool in the socio-economic development of a nation. These services provide necessary support for the rapid growth and modernisation of various sectors of the economy. By December 2011, telephone subscriptions in India totalled 951.34 million. The overall tele-density in the country was 78.66, with the rural tele-density at 39.22.[5] This telecom subscriber base is supported by around 400,000 towers, with another 100,000 on the way. At present DG sets with a capacity of 10–15 kVA are used as an alternative source of power in thousands of base transceiver stations (BTSs).[6] Often, BTSs with DG sets are operated for the entire day even in the presence of grid supply, simply to avoid voltage fluctuations that could disrupt services and destroy the expensive and sophisticated equipment. On an average, the load consumption of a BTS is about 8 kW. The Greenpeace report 'Enabling Green Talking' (August 2012) estimates that 3.2 billion litres of diesel were consumed by the telecom industry in 2011 and that consumption is expected to reach 6 billion litres by 2020. Renewable energy has the potential to provide a cost-competitive alternative to DG sets, particularly in rural areas. Two options are possible. The more common one is to use solar power during the day. The other is biomass gasification or biogas-based option. Experiments are also being conducted to have a small dedicated 10 kW gasifier that could provide 24-hour service. A couple of thousand towers are indeed solar-powered, some supported by a subsidy for a few hundred towers by the Ministry of New and Renewable Energy in 2010. However, it is expected that this subsidy will not be required for long because this is a viable proposition in itself and is becoming more so as time goes on. It is unfortunate that tower companies have been delaying investment to power the towers through renewable energy means and thereby saving themselves and the country from huge and unnecessary diesel consumption.

In March 2013, the Government of India approved the recommendations put forth by the Telecom Regulatory Authority of India on Green Energy Applications (TRAI, 2011). It inter alia provides for at least 50 per cent of all rural towers and 20 per cent of urban towers to be powered by hybrid power, i.e., renewable energy technologies and grid power by 2015, while 75 per cent of rural towers and 33 per cent of urban towers are to be powered by hybrid power by 2020.

There are two measures that directly relate to energy access. The first is what we discussed in Chapter 6 in the models with the attempt of some companies to use power for telecom, where they can afford to pay much more per unit, as the anchor load. The second is that the use of solar power in large quantities would help develop the solar ecosystem at the local level. Ideally, both measures should proceed in tandem. Energy systems for telecom towers and villages could complement each other.

Rural Health Centres

Primary healthcare in rural India essentially comes from around 25,000 primary health centres (PHCs) and 150,000 sub-centres. Broad estimates indicate that the average load requirement for a PHC is around 2 kW, and for a sub-centre, around 170 W. With healthcare facilities expected to increase, the demand for small-power generation will increase in the long term. A large number of these centres do not have electricity—or if they do, it is unreliable and unpredictable. As discussed in Chapter 3, rural areas are likely to remain deprived of grid electricity for a long time. The public places in these areas, particularly health sub-centres in many parts of the country, are not likely to get electricity. The same is likely to be true for Anganwadi centres, where small children are supposed to gather for their morning meal as well as immunisations. These centres are intended also as venues for preschool education as well as gathering places for pregnant mothers. A simple light in these rooms would improve the atmosphere significantly. And in many places solar-powered electricity may be the only way to make toilets functional with running water. Therefore, solar energy can easily provide the basic needs for such centres. In fact, it is necessary, because success in efforts towards reducing infant and maternal mortality and malnutrition are dependent upon the proper functioning of these centres. As far as sub-centres are concerned, availability of solar-based electricity may allow auxiliary nurse midwives (ANMs) to also stay at their headquarters, at least in the longer term. Where PHCs are unable to function properly because of the absence of light and electricity for certain critical appliances, solar energy can make all the difference. In J&K, especially in Ladakh, many PHCs have been provided with rooftop solar systems. In Chhattisgarh, all PHCs and community health centres (CHCs) have been provided with solar power. They have also been given solar-powered drinking water facilities. A comprehensive five-year programme to solarise all such institutions in the country, in areas where these are needed, would have a great impact.

Rural Banking

According to the 2011 Census, there are around 100,000 branches of Scheduled Commercial Banks in the country, of which around 36,000 are in rural areas. Still, a significant proportion of rural households are outside the formal fold of the banking system, and this segment is expected to grow over the years. The plan for transferring subsidy benefits by cash means that branches and transactions will increase manifold. Mobile banking may be developed, and ATMs may also be set up. A village-based bank branch, usually operational only during the day, requires about 1.5 kW of power. Hundreds of branches use diesel generation as their power source today. In 2010, the Ministry of New and Renewable Energy suggested to the Finance Ministry a project to power all rural bank branches with solar energy and have this project funded through the National Clean Energy Fund. This could serve as a wonderful demonstration for visitors to the bank and could be used as an example by other public and non-public institutions functioning in those areas, including small business enterprises. The bank personnel themselves could become votaries of solar power and encourage or promote lending for solar home lights—for households in general and affluent ones in particular.

Rural Education

The problems of schools in rural areas, particularly away from the main towns, are well known. Buildings are poorly constructed and have rooms that remain dark even during the day, especially when the small windows have to be closed to prevent the heat from coming in. Children sit on the floor in rows with barely enough light to look at their books or write. Good lighting during the daytime will make a huge difference, as will fans. Are rural children not entitled to such minimal comforts? In thousands of schools, lights and fans will make a huge difference in encouraging attendance and facilitating greater attention. The second problem is the quality of teachers. The manner in which teachers are recruited does not give confidence that rural schools will have suitable personnel to teach the fundamentals of languages, science and mathematics. This is a well-debated issue in India. While schools have proliferated and enrolment has increased, the focus now is on learning levels. There are existing TV educational programmes but they are made primarily for urban children. Additional ones need to be made for children in rural areas, in their native languages, and relevant to their living and learning environment. This is a huge task: hundreds of lectures will have to be prepared in different languages and even some dialects. This should immediately be done in mission mode. The technology exists, and solar power will enable such tutorials to be communicated to children via TVs/computers in school buildings that have lights and fans. This could mean evening classes also. A major problem in rural schools is the absence of teachers or the lack of teaching skills. Therefore, these schools need a huge programme to teach through the distance mode. Teachers would need to be trained to supplement the TV lectures. A village environment with lights in houses and streets and a vibrant school will create an altogether different and positive atmosphere, the results of which may have revolutionary consequences. It is also well established that the most immediate benefit of proper lighting in homes is that children get more time to study.

It is now being increasingly recognised that all schools must have toilets. Previous experience suggests that while toilets have been built, they have not been properly used because of lack of functionality in the absence of running water. In Chhattisgarh, girls' schools are provided with drinking water facilities. Prime Minister Modi has called for universal coverage of toilets in schools. Solar energy may be the best way to ensure functional water supply systems and toilets in schools. This can be done everywhere.

Public Places and Service Centres

The RGGVY evaluation and the 2012 Greenpeace report clearly show that while records or statistics indicate that most public places in rural areas have energy access, the fact is that large parts of the country either have no electricity connections or have connections with little actual supply of power and have to resort to diesel sets, lanterns, etc. The environment would change drastically if all offices had fans, lights and computers powered by solar energy during the day and the houses of officers and staff were powered by solar energy during the evenings. Rooftop solar installations in many police stations in the naxal affected areas of Chhattisgarh have made a lot of difference.[7]

At one time, the Government of India came up with the idea of opening common service centres in villages; these centres would have computers and other equipment to help provide basic services such as photocopying and making reservations. In 2010 there were several discussions to power these, where required, through solar energy. But it did not happen. There is a need to revisit this proposal. The Government of Rajasthan has provided such centres and facilities in all panchayat offices of the state. As the years go by, this will enhance work ethos and service delivery.

Agricultural Pump Sets

This brings us to the revolutionary possibilities of solar water pumping. Agriculture in India contributes to 15 per cent of the GDP and employs 50 per cent of the country's workforce. Reliable irrigation, as we know, is a critical requirement for the farmer. With the increase in local demand for food grains and other agricultural products, there has been a considerable increase in the number of pump and pump sets[8] sold in India. Such pumps/sets are essential for irrigation and cultivation, as a large proportion of the fertile land in India falls within the dry belt region of the country. In these areas, around 80 per cent of the available water is being used for agricultural purposes, covering over 80 million hectares of arable land. Water is drawn from deep wells and shallow tube-wells. As a result, India today has around 18 million grid-connected pump sets and 7 million diesel-operated pump sets. However, erratic grid supply and the high cost of diesel pumping continue to remain problems for farmers. In many areas, and sometimes for many days, irrigation via electric pumps is possible only at night, due to availability of electricity during that time only. Poor irrigation as a result of these issues results in significant yield losses. While a scientific assessment on yield loss due to poor quality of irrigation has not been done, it is generally understood that crop yields will improve if the appropriate volume of water is available when required.[9] In addition, in Bihar and other areas where water tables are high, there is likely to be additional cropping during the summer season.

KPMG carried out a brief study on the economics of solar irrigation and suggested a plan of action, which is summarised below (Mohit, 2013).

- Current economics indicate an internal rate of return (IRR) of 10 per cent for replacement of diesel pumps with solar pumps. which improves to 19 per cent when yield benefits are considered
- Upfront cost of a solar pump (2.2 kW) is about 10 times of a conventional pump (₹3.5 lakh vs. ₹30,000). Capital subsidy level required today is around 70 per cent (₹2.5 lakh for a 2.2 kW pump). This can come down to 55–60 per cent if bank financing support is made available.
- With a market support programme, costs can down by 25–30 per cent over the next 4–5 years and a financing ecosystem can develop. The capital subsidy requirement can then come down to ₹70,000 per 2.2 kW pump compared to ₹2.5 lakh today.
- A government-supported market assurance programme has been recommended in 12th and 13th Plan periods—2 lakh pumps (50,000 per year) in the 12th Plan and 8 lakh pumps in the 13th Plan. The first 1 lakh pumps would require a capital subsidy outlay of ₹2,500 crore, and the second with reduced subsidy of 60 per cent a capital subsidy ₹2,150 crore, totalling ₹4,650 crore for 2 years. During the first half of the 13th Plan, i.e., for the next 4 lakh pumps, a reduction in subsidy to 50 per cent is recommended. Thereafter, the government may assess the prevailing market situation and re-evaluate subsidy requirement for the remaining half of the 13th Five Year Plan.

The benefits to the government (and country) are as follows:

- Replacement of 1 million diesel pumps with solar pumps would result in diesel use mitigation of 9.4 billion litres over the life cycle of solar pumps and a diesel subsidy saving of ₹8,400 crore.

- Forex savings of US$300 million per annum on diesel imports for replacement of 1 million diesel pumps translating into forex savings of US$4.5 billion over pump life.
- Better crop yields that can improve agricultural output by ₹2,000 crore per annum or ₹30,000 crore over the pump life, or even more.

This is an eminently feasible programme. The calculations may need to be scientifically verified, but in general the figures can be accepted for the purposes of a policy decision. Studies should be done regarding yield improvement where pumps have actually been installed, and some more pilot studies should be carried out with that specific purpose in mind. The full benefits need to be properly understood. The above emphasises only the partial benefits in terms of yield improvement and diesel savings. But the perspective is much larger. Yield improvements, including unassessed additional summer crop yields in many areas of Bihar, will considerably increase incomes of farmers as well as provide additional rural employment, including in the lean summer season. If solar pumps are installed where currently there are no pumps—electric or diesel—the above benefits will also accrue. In Chhattisgarh, where many applications for electric pumps are pending, this is already happening. Moreover, there is considerable scope to provide power to many lift irrigation schemes lying defunct due to problems with electricity supply. This is being planned in Odisha. If the subsidy instrument is better designed, marginal farmers can be targeted. This will have the added benefit of poverty alleviation. Once costs have been reduced and the ecosystem established, there can be 10–20 million pumps in the next couple of decades. A properly designed banking instrument and funding from the National Clean Energy Fund will finance this effort.

There are two additional benefits that could have very important consequences. First, very low rates for power for irrigation have had adverse impacts on the finances of utilities. Any shift from grid electricity to solar power would help these utilities considerably, leading to improvements in the health of utilities as electric pumping is hopefully replaced in many areas by solar-powered pumping. This could be one of the best power demand management measures. Needless to say, the same will be the impact of the mini-grid. Second, this will help to develop the solar ecosystem in rural areas as well as create a huge market for solar. Actually all the different actions proposed, including electricity access, would do the same. Faster the development of this ecosystem, better and faster would be further solar deployment. It would then be a virtuous cycle. It is essential, therefore, to announce a large programme for installation of solar pumps as proposed above and earmark resources while working out the details of location, clustering, identification of beneficiaries, etc. Solar pumping will be the way to the next agricultural revolution, and certainly so for Bihar and similar areas as we move towards eastern India.

We can extend the concept of pumping water through solar pumps to provision of village drinking water supplies. The process started by giving funds to the drinking water mission from the National Clean Energy Fund. The result needs to be studied for large-scale up-scaling.

Impact

It is not difficult to visualise the benefits of the above measures along with the benefits we have already discussed: considerable health benefits with the provision of biogas or improved cooking stoves, as well as saved healthcare costs; better education and environment in schools; better health facilities; more working days in which to earn a livelihood; saved fuel collection and cooking time for women; better hygiene and sanitation in villages; provision of drinking water facilities, better lighting in the evenings, which lead to enhanced socialising and liveliness; possible sources of private and public entertainment through TV; shops open until late in the evening; additional agricultural production and increased incomes; local employment and entrepreneurial generation. After a review of several programmes across the world, it becomes clear that once an area has reached a certain level of development, further progress in raising standards of living to socially acceptable levels depends upon the availability of electricity supply. But we would argue that in the Indian context, particularly in the most backward regions, energy access would be a necessary precondition even to reach that minimum level of development. Unquestionably, therefore, the provision of energy access must become one of the most important policy priorities of the government.

Notes

1. CRISIL Research Insight (August 2012). Sustaining the rural consumption boom. Available at http://www.crisil.com/pdf/economy/research-insight_rural-consumption_Aug12.pdf. Accessed on 23 November 2014.
2. Fourth All India Census of Micro, Small and Medium Enterprises 2006–07. Available at http://www.dcmsme.gov.in/publications/Final%20Report%20of%20Fourth%20All%20India%20Census%20of%20MSME%20Unregistered%20Sector%202006-07.pdf. Accessed on 23 November 2014.
3. 'Integrated Electricity Supply Master-Plan for Off-grid Supply in Sundarban Islands', TERI Report, 2005.
4. The markets operate mostly during the day and in some cases in the evening hours. In order to calculate market potential, the entire village market segment has been considered to operate during the day as well as in the evening hours. In addition, there is the potential for small commercial markets in semi-urban and urban settings.
5. TRAI Annual Report 2011–12. Available at http://trai.gov.in/WriteReadData/Miscelleneus/Document/201301150318386780062Annual%20Report%20English%202012.pdf. Accessed on 23 November 2014.
6. Interviews with Reliance Infocomm Ltd (mobile services provider), New Delhi.
7. Based on discussions with state officials.
8. A pump set refers to a suction valve connected to a diesel engine.
9. The International Water Management Institute and TATA Water Policy Research Program have been assessing the benefits of solar water pumping in Bihar.

10
The Last Word

In the first chapter, we discussed the importance of energy access for human development of somewhat marginalised populations in rural areas. In the previous chapter, we have discussed the possible impact on rural development in India. Our attempt is to place energy access into the larger picture of empowerment and of equitable and inclusive growth. Energy access is hardly discussed in the above context. We are not economists, so we rely on opinions of well-known economists and experts to briefly discuss this perspective.

Achieving equity is a major concern. Article 38(2) of the Constitution states that 'the State shall, in particular, strive to minimise the inequities in income and endeavour to eliminate inequalities in status, facilities and opportunities'. In the 12th Plan document, the issue of equity in development has been briefly stated thus:

> some increase in inequality in a developing country during a period of rapid growth and transformation may be unavoidable and may even be tolerated if it is accompanied by sufficiently rapid improvement in the living standards of the poor. However, an increase in inequity with little or no improvement in the living standards of the poor is a recipe for social tensions. Static measures of inequality do not capture the phenomenon of equality of opportunity which needs special attention. Any given level of inequality of outcome is much more socially acceptable if it results from a system which provides greater equality of opportunity.[1]

There cannot, therefore, be any doubt that the achievement of high growth must ultimately be judged in terms of the impact of that economic growth on the lives and freedoms of all people. That is why it is important that India address issues of equity comprehensively.

In the two decades since India embarked on economic reforms, there has been tremendous growth and great gains in poverty reduction. Poverty declined to 21.9 per cent in March 2012 from 37.2 per cent in 2004–05. There is general consensus that 'extreme poverty, long considered an immutable fact of life in India, is finally in retreat' (Gupta et al., 2014). But millions of Indians continue to face significant deprivation in terms of quality of life and access to basic services. How does one assess the impact and extent of entrenched energy poverty, which is very much there? This could be done partly by proxy indicators represented by the index of multi-dimensional poverty and the empowerment line proposed by the Mckinsey Global Institute. This line identifies a notional line below which living conditions continue to remain quite abject. Estimated in financial terms at ₹1,544 per capita per month it is more than 50 per cent higher than the official poverty line. As of 2012, 680 million people, considered the deprivation number, had consumption levels below this line. Of these, 57 million are considered 'excluded', which are the poorest of the poor; 210 million are 'impoverished' and 413 million 'vulnerable', which means that they have a tenuous grip on a better standard of life. In 2005–06, the multi-dimensional poverty index was 48.5 per cent of India's population as against 37 per cent of the officially poor. Today when the officially poor have declined to 22 per cent, this index covers 55.3 per cent of the population![2] These are the people whose lives energy access, and other social determinants, will touch and improve.

We had concluded Chapter 1 by emphasising the problems of the more backward areas of the country. The Mckinsey Study gives the highest access deprivation scores as follows: Bihar—62; Uttar Pradesh—57; Jharkhand—54; Madhya Pradesh—49; Assam—48 and Odisha—47. These are the problem areas from the point of view of energy access also. More specifically, the Report finds, not surprisingly, that India's most deprived districts are heavily concentrated in Uttar Pradesh and Bihar. And their deprivation scores of energy stand at 82 per cent, which are considerably higher than the national average of 59 per cent. Clearly, therefore, provision of energy access in these areas could make a dramatic impact on peoples' lives.

During the last decade, while India has been climbing up the ladder of per capita income it has unfortunately been slipping down the slope of key social indicators (in comparison with others, including our neighbours in South Asia). It is a shame that we are ranked 135th in the Human Development Index in 2014. In fact, Dreze and Sen point out rather grimly that 'the history of world development offers few other examples, if any, of an economy growing so fast for so long with such limited results in terms of reducing human deprivation' (Dreze and Sen, 2013). Although the issue of inequality is complex, and there could be differing interpretations or conclusions, there is a general perception that inequality is actually substantively and demonstratively increasing, and it is manifested in many different ways. Gini co-efficient between 2004 and 2012 rose from 0.27 to 0.28 in rural areas (Jha, 2013). The rural urban divide is getting bigger, especially in terms of availability of opportunities. Our growth story is reflecting the Great Gatsby curve defined by Alan Krueger, which shows that as income inequalities grow, opportunities for upward social progression reduce. Richer get richer and others struggle to stay on the ladder. The President of the IMF, Christine Lagarde, in a public lecture stated that 'in India, the net worth of the billionaires' community increased 12 fold in 15 years, enough to eliminate absolute poverty in this country twice over'.[3] While energy access concerns the bottom of the Pyramid, a Report which made headlines in July 2014 in India talked about the Top of the Pyramid.[4] It claimed India's ultra-rich club is growing fast. The net worth of ultra-high net worth individuals (UHNIs—net worth above 25 crores, currently 117,000 individuals) is projected to surge at an annual compounded rate of 34 per cent from an estimated ₹104 lakh crore in FY 2013–14 to ₹408 lakh crore by FY 2018–19. It is these inequalities that have led Amartya Sen and Jean Dreze to argue that it is not only that the new income generated by economic growth has been very unequally shared but also that 'the resources newly created have not been utilised adequately to relieve the gigantic social deprivations of the underdogs of society making the country look more and more like the islands of California in a sea of sub-Saharan Africa' (Dreze and Sen, 2013). It is, therefore, sometimes said that in India there has been growth but not adequate development. And surely, there must be something wrong with our developmental or growth model, which someone described somewhat harshly as that of 'crass consumerism, crony capitalism, cancerous corruption and increasing divisions between India and Bharat' (Kapur, 2012). It is time that we come to terms with this reality and look for a new course, a new identity and a new model which combines growth with genuine inclusiveness, justice, equity and sustainability. Inclusiveness is not just about bringing those below an official fixed poverty line to a level above it. It is about a growth process seen to be fair; it embraces concerns about certain marginalised groups (with women being seen as one such group), greater attention to income equality, urban–rural consumptions. It is not only about ensuring a broad-based flow of benefits or economic opportunities but also about empowerment. If there is such change, universalisation of energy access would become one of the principal priorities for state policy, and a principal measure to judge its success.

It is obvious that we need growth, which requires bold measures. But we must also address inequality and inclusiveness. Growth will give us trickle down as well as generate resources to tackle inequitable conditions. The question is how to tackle these conditions. There are important differences of approach. One approach was to focus more on expansive populist schemes like the expanded Food Security or MGNREGA. India has substantially raised expenditure on social welfare programmes since 2005. But the impact has been muted—partly because, as we shall see, large parts of funds did not reach the intended beneficiaries. Others feel that the better instruments would be the social determinants like access to energy, clean water, sanitation, health and education. Considerably more resources have to be invested in social infrastructure and access to these basic services, and over a sustained period of time. No doubt it would be necessary also to ensure efficient and effective delivery of these services. We completely agree with this view. As we have seen, addressing the severe shortcomings in these areas would be critical in solving India's poverty challenge. Both approaches are necessary.

The problems with these schemes are that they take away huge resources, reach undeserving populations and have huge leakages. There is a general public consensus on these issues. Economist Surjit Bhalla has discussed the issue of leakages in the subsidies on MGNREGA and Food Security. In 2011–12, the difference between what food grain value the government said it delivered and what people actually received was ₹30,000 crore. Under MGNREGA, ₹34,000 crore was received by non-poor, and out of ₹105,000 crore spent on MGNREGA/food subsidies, only 13 per cent went to the deserving poor. In that year, per capita expenditure of poor (BPL) was ₹709 with the poverty line being ₹893.

About ₹55,000 crore was needed to bring them to the poverty line. We reduced only 40 million, which could have been done by ₹8,000 crore (Bhalla, 2013). He has discussed a similar pattern in earlier years where he concluded that such direct measures have been singularly ineffective in generating inclusive growth outcomes (Bhalla, 2011). The Mckinsey study also concludes that in 2009–10, 35 per cent of the food subsidy did not reach consumers and only 36 per cent of it reached BPL beneficiaries. Only 52 per cent of MGNREGA and 47 per cent of fuel subsidy reached intended beneficiaries.

Economists could debate the maths of the above figures, but the message is clear. It is better to create capabilities in practice rather than rights in theory. Should we address human deprivation directly or through human development? Should we concentrate on entitlements in theory or realise them in practice through inclusive growth? Should we continue to spend excessively huge amounts on direct interventions or in simultaneously addressing the circumstances that lead to these problems and giving substantial resources for these? There has been a raging public debate on what constitutes the poverty line, but hardly any discussion on the more relevant debate regarding the discussion on what is our HDI status and achievements in key MDG indicators (including energy access) and how they can be improved. The huge expenditures on these schemes also show that the availability of resources is not the real problem, only there is over-emphasis. Since 2007–08, there has been a massive shift from capital spending to subsidy. The latter are up from 1.5 to 2 per cent of GDP while the former has declined from 2.5 to 1.75 per cent of GDP. All this is in the name of inclusiveness—inclusiveness means more jobs and not more handouts (Ninan, 2014). Redirect the subsidies and concentrate on development to ameliorate the causal factors for inequality by looking at a broader range of solutions rather than make huge (and leaky) direct interventions. As Arvind Panagariya, Professor of Political Economy at Columbia University, now Deputy Chairman of the newly set up Niti Ayog, says, we must not confuse inclusive growth with inclusive spending (Panagariya, 2013).

The current approach highlights two further consequences. First, as Sen and Dreze say, 'we also have to recognise with clarity that the neglect—or minimising—of the problem in public reasoning is tremendously costly, since democratic rectification depends crucially on public understanding and widespread discussion of the serious problems that have to be addressed'. All through this book we have tried to show the neglect of the real problems of energy and energy access. This has hindered development of policies which are needed to address them. The relative neglect of social indicators have the same consequences, otherwise we would have given much more resources and effort to reduce malnourishment and infant/maternal mortality and provide drinking water, sanitation, energy access, etc.

The second consequence, they argue, is that the biases of public policy towards privileged interests also take many other forms, including the neglect of agriculture and rural development, and the showering of public subsidies (implicit or explicit) on privileged groups. Recently, IMF mentioned that India's richest get 6 times the subsidy of 20 per cent of India's poorest! In our specific context, a noted economist (now India's new Chief Economic Advisor) has argued that the worst economic policy in India is the energy subsidies for diesel, kerosene and above all power. Consider, he says, the bad outcomes that power subsidies cause or abet—bad crop mix; depleted water resource; unprofitable and mismanaged State Electricity Boards; under investment in power, lower economic growth and higher carbon emissions (Subramaniam, 2013). And to these we add lack of energy access. We have discussed the issue of subsidies in some detail and provided enough evidence that energy access can be adequately financed through a simple redirection of subsidies and by part of the cess on coal.

In the larger context of energy requirements and developmental models, there is a global worry about the huge unmet energy needs of India's vast population. They have seen the impact of China's huge growth in energy generation and consumption, which has become the world's biggest carbon emitter. In fact, India's lesser energy consumption has provided the world with breathing space. What if these needs get tied to fossil fuels? It has, therefore, been argued that India is actually on the horns of an 'energy trilemma'—energy security, equitable energy access and environmental impact mitigation (Neil and Thomas, 2012). The provision of energy security remains a principal challenge to, and an imperative for, India's economic growth aspirations and its efforts to improve human development. Clearly, a path less dependent upon fossil fuels for energy is not only desirable but has virtually become a necessity and a drive towards adoption of more renewable energy seems inescapable. It is also increasingly being recognised, as the Asian Development Bank Report notes, that economies that go green fastest will be the most prosperous 20–30 years from now. Nations with the fewest energy imports, lowest subsidies, most innovative renewable energy sources and foreign policies that are not beholden to oil suppliers will be most attractive and have the best credit ratings. Structurally, this may well be easier for countries such as ours. Since fossil-based energy infrastructure is not fully developed, it is possible that we may be able to manage transition to renewable energy more easily and there may be competitive advantages in doing so. In developmental jargon this may be called 'leap frogging'. Nowhere is this picture clearer or more evident than in promoting renewable energy solutions for rural electrification. And when we talk of energy security of the nation, should we leave out the energy security of the individual, the poor, and those residing in rural areas? For India, therefore, there is a larger role for renewable energy than merely going clean or reducing the carbon footprint. It will be a multi-pronged tool to meet shortage of power from conventional sources; reduce consumption of fossil fuels; mitigate carbon emissions; improve all kinds of deficits (current account, budget, fiscal) by savings on imports and elimination or reduction of subsidies; lead to energy and electricity demand management; become an instrument to improve the financial health of utilities while reducing future tariffs; and provide energy access in the broadest terms with all its huge positive economic and social externalities.

Energy access policies cannot be framed in isolation to the nation's economic development policies. The overall synergy can address a broad range of aims, including energy security and equity issues. Further, the cascading economic social and environmental benefits with households gaining access to modern energy would include improvement in the standards of living, which will lead to pay for higher amounts and costs of energy in future. There is universal agreement that absence of access to energy leads to low economic activity and thus low demand, which turn off energy utilities to extend energy services—in a way, a vicious cycle. Lack of affordability and erratic supplies have limited ability to use resources. As we have discussed in the book, barriers are multi-dimensional and there are many interlinking issues. On the policy front, therefore, our approach must also be multi-dimensional in nature to overcome these barriers.

Universal energy access challenges are gigantic and exciting. There is no magic bullet solution but the potential of decentralised renewable energy-based solutions is enormous. The solutions will have to be embedded within the national development plan and strategy and would certainly need support from comity of nations under the existing multilateral and bilateral framework both for finances and technologies.

As we examine global developments in the energy sector, it is clear that a new paradigm is developing as the design criteria for energy systems change—from a centralised, inflexible, commodity-like system to a substantially decentralised, flexible, modular, which is integrated with multiple levels and manners of service delivery. With it will come a new developmental model and a different lifestyle with supporting structures. There are three possible future systems—fossil fuel centric; high nuclear-coal centric and high renewables-highly distributed-energy efficient-low demand centric.[5] The last model will clearly win. In the long run this seems to be our only chance. The policy makers need to be part of the change, adapting and changing the policies to meet new design requirements made necessary by technological developments. This is the real revolution that the idea of renewable energy offers. Jeffery Rifkin in his book 'The Hydrogen Economy' has rightly envisioned the potential of renewable energy-based distribution-generation where people can produce renewable energy and share it peer-to-peer, just like they now produce and share information, creating a new, decentralised form of energy use. This is the way forward for India to achieve its goal of universal energy access.

Notes

1. Planning Commission, 12th Five Year Plan, Chapter 1, Vol. 1.
2. UNDP Human Development Report 2014.
3. Quoted in *Indian Express*, 24 April 2014.
4. Top of the Pyramid, 2014, Kotak Wealth Management. Available at www.kotak.com/sites/default/.../top_of_the_pyramid_2014_kwm.pdf. Accessed on 23 November 2014.
5. Renewable Energy: Global Futures Report, REN 21, Paris, 2013. Available at.http://www.ren21.net/portals/0/documents/resources/gsr/2013/gsr2013_lowres.pdf. Accessed on 23 November 2014.

Bibliography

Adams, M. (2014). India's energy future: The EIU view. *The Economist*, a report from the Economist Intelligence Unit. Available at: http://www.economistinsights.com/sites/default/files/downloads/Empowering_Growth.pdf. Accessed on 22 November 2014.

Ahluwalia, M.S. (2011). Prospects and Policy changes in the Twelfth Plan. Available at www.planning commission.nic.in/about us/speech/spemsa/spe_21052011.pdf. Accessed on 22 November 2014.

Ahn, Sun-Joo and Dagmar, G. (2012). Understanding energy challenges in India: Policies, players, and issues. International Energy Agency. Available at http://www.iea.org/publications/freepublications/publication/India_study_FINAL_WEB.pdf. Accessed on 22 November 2014.

Anand, S. and Sen, A. (1997). Concepts of human development and poverty: A multi-dimensional perspective. In *Poverty and Human Development*: Human Development Papers. New York: UNDP.

Alkire, S. and Santos, M.E. (July 2010). Multidimensional poverty index, Oxford Poverty and Human Development Initiative.

Banerjee, R. (2006). Comparison of options for distributed generation in India. *Energy Policy*, 34(1), 101–11.

Barnes, D. and Foley, G. (2004). Rural electrification in the developing world: A summary of lessons from successful programs: Joint UNDP/World Bank Energy Sector Management Assistance Programme (ESMAP), World Bank, Washington DC: http://iis-db.stanford.edu/evnts/3961/Doug_Barnes_paper.pdf. Accessed on 23 November 2014.

Barnes, D.F. and Sen, M. (2002). Energy strategies for rural India: Evidence from six States. ESMAP/World Bank: http://imagebank. worldbank.org/servlet/WDS_IBank_Servlet. Accessed on 23 November 2014.

Barnes, D.F., Kumar, P. and Openshaw, K. (2012). *Cleaner Hearths, Better Homes. New Stoves for India and the Developing World*. New Delhi: Oxford University Press.

Batliwala, S. and Reddy, Amulya K.N. (1996). Energy for women and women for energy: A proposal for women's energy entrepreneurship. Available at http://www.esmap.org/sites/esmap.org/files/Energy%20for%20Women%20and%20Women%20for%20Energy.pdf. Accessed on 23 November 2014.

Bhagwati, J. and Panagariya, A. (2012). *India's Tryst With Destiny: Debnking Myths That Undermine Progress and Addressing New Challenges*. New Delhi: Harper Collins.

Bhalla, S. S. (2011). Inclusion and growth in India: Some facts, some conclusions, London School of Economics Asia Research Centre. Working Paper Number 39.

———. (2013). Rajiv Gandhi lessons on the Food Bill. *Indian Express*, 27 July.

Bhattacharyya, S.C. (2011). Review of alternative methodologies for analysing off grid electricity supply. OASYS South Asia Project.

Boyle, G. and Krishnamurthy, A. (2011). Taking Charge: Case studies of decentralized renewable energy projects in India in 2010. Greenpeace (p. 58). Available at http://www.greenpeace.org/india/Global/india/report/2011/Taking%20Charge.pdf. Accessed on 22 November 2014.

BP Statistical Review of World Energy. 2013. Available at https://www.bp.com/content/dam/bp/pdf/statistical-review/statistical_review_of_world_energy_2013.pdf. Accessed on 22 November 2014.

Bruce, N.G., Relifuess, E.A. and Smith, K.R. (2011). Household energy solutions in developing countries. In Nriagu, J.O. (ed.) *Encyclopedia of Environmental Health*, vol. 3, pp. 6275. Burlington: Elsevier.

Central Electricity Authority. (2012). All India Electricity Statistics, General Review.

Census of India. (2011). Office of the Registrar General & Census Commissioner, India.

Centre for Science and Environment. (2012). Going Remote: Reinventing the off-grid solar revolution for clean energy for all. Available at http://shaktifoundation.in/wp-content/uploads/2014/02/going%20remote.pdf. Accessed on 22 November 2014.

Chatterjee, V. (2013). Shrivelling temples of resurgent India. *Business Standards*, 19 August 2013. Available at http://www.business-standard.com/article/opinion/shrivelling-temples-of-resurgent-india-113081901183_1.html. Accessed on 22 November 2014.

Chaurey, A. and Kandpal, T.C. (2009). Carbon abatement potential of solar home systems in India and their cost reduction due to carbon finance. *Energy Policy*, 37, 115–25.

———. (2010). A techno economic comparison of rural electrification based on solar home systems and PV microgrids. *Energy Policy*, 38(6), 3118–29.

———. (2010). Assessment and evaluation of PV based decentralized rural electrification: An overview. *Renewable and Sustainable Energy Reviews*, 14(8), 2266–78.

Cust, J., Singh, A. and Neuhoff, K. (2007). Rural electrification in India: Economic and institutional aspects of renewables. Available at http://www.eprg.group.cam.ac.uk/wp-content/uploads/2014/01/eprg0730.pdf. Accessed on 23 November 2014.

De, S. (2013). Developing modern India. *Energy Next*, 3(9), 17–19.

Deorah, S. and Chandran-Wadia, L. (2013). Solar mini-grids for rural electrification: Observer Research Foundation, Mumbai. Available at http://www.google.co.in/url?sa=t&rct=j&q=&esrc=s&source=web&cd=1&ved=0CCgQFjAA&url=http%3A%2F%2Fwww.arthaplatform.com%2Fdownload%2F%3FKey%3DZ02ZbQ47LJHP5Ade-DQ1PqLlFH%2Bnyek14unyvObg8U3yLHGFdQKA7fMFGHjXJAxr3Xe0I11L2inKbhYSZImkOC6D4xkphqwnErRpBRJ2pYovk%3D&ei=QOyFUu7dJcTTrQfD_YCYBw&usg=AFQjCNGwKFhWT3yNtRP3w4jW6DGViUTy7A&bvm=bv.56643336,d.bmk. Accessed on 22 November 2014.

Dreze, J. and Sen, A. (2013). *An Uncertain Glory: India and Its Contradictions*. New Delhi: Penguin.

Ekouevi, K. and Tuntivate, V. (2013). Household energy access for cooking and heating: Lessons learned and the way forward. World Bank's Energy and Mining Sector Board Discussion Paper No 23. Available at http://siteresources.worldbank.org/EXTENERGY2/Resources/HouseHold_Energy_Access_DP_23.pdf. Accessed on 23 November 2014.

Electricity Act 2003: Ministry of Power (2003), Government of India.

Eighteenth Electric Power Survey. (2013). Central Electricity Authority, Government of India.

Eckholm, E. (1975). 'The Other Energy Crisis: Firewood', Worldwatch Paper No.1, Worldwatch Institute, Washington, D.C.

Faruqui, A. et al. (2010). The impact of informational feedback on energy consumption—A survey of the experimental evidence. *Energy*, 35(4), 1598–1608. Abstract available

Bibliography 251

at http://www.sciencedirect.com/science/article/pii/S0360544209003387. Accessed on 22 November 2014.

Financial Express. (4 April 2014). Rural India outpaces urban spending. Available at http://www.financialexpress.com/news/rural-india-outpaces-urban-spending-/995115. Accessed on 23 November 2014.

Foley, G. (1995). Photovoltaic applications in rural areas of the developing world. World Bank Technical Paper, 304, Washington DC: http://www-wds.worldbank.org/servlet/WDSContentServer/WDSP/IB/1999/08/15/000009265_3961214155110/Rendered/PDF/multi0page.pdf. Accessed on 23 November 2014.

G-8 Task Force on Renewable Energy (2001). Available at http://www.climate.org/PDF/g8_ren_energy.pdf. Accessed on 23 November 2014.

Gadgil, M. et al. (2012). Report of the Western Ghats Ecology Expert Panel. Ministry of Environment and Forests, Government of India. Available at http://www.moef.nic.in/downloads/public-information/wg-23052012.pdf. Accessed on 22 November 2014.

Ghosh, A., Majumdar, S. and Kadam, G. (2012). State-owned electricity distribution companies: Some positives, though several concerns remain. Available at http://icra.in/Files/ticker/Power%20Distribution%20Note.pdf. Accessed on 22 November 2014.

Government of India. (2005). National Electricity Policy: Ministry of Power. Available at http://www.powermin.nic.in/whats_new/national_electricity_policy.htm. Accessed on 22 November 2014.

———. (2006). National Mission on Decentralized Biomass Energy for Village Industries. Unpublished document submitted to Secretariat of the National Advisory Council in January 2006.

———. (2006). National Rural Electrification Policy: Ministry of Power. Available at http://powermin.nic.in/whats_new/pdf/RE%20Policy.pdf. Accessed on 23 September 2014.

———. (2008a). National Hydro Power Policy: Ministry of Power. Available at http://www.ielrc.org/content/e0820.pdf. Accessed on 23 September 2014.

———. (2008b). National Action Plan on Climate Change. Prime Minister's Office. Available at http://www.c2es.org/international/key-country-policies/india/climate-plan-summary. Accessed on 22 November 2014.

———. (2012). Twelfth Plan documents. Planning Commission, Government of India. Available at http://planningcommission.nic.in/plans/planrel/fiveyr/welcome.html. Accessed on 22 November 2014.

Granado, J.R. del, Coady, D. and Gillingham, R. (2010). The unequal benefits of fuel subsidies: A review of evidence for developing countries. IMF Working Paper No. WP/10/202. Available at http://www.imf.org/external/pubs/ft/wp/2010/wp10202.pdf. Accessed on 23 November 2014.

Gupta, R., Sankhe, S., Dobbs, R., Woetzel, J., Madgavkar, A. and Hasyagar, A. (2014). From poverty to empowerment: India's imperatives for jobs, growth and effective basic services. Mckinsey Global Institute. Available at http://www.icrier.org/pdf/mgi_poverty_v2.pdf. Accessed on 23 November 2014.

Ishofsky, R. (2013). Will Obama's power Africa initiative reach the rural poor? *Forbes,* 20 August, 2013. Available at http://www.forbes.com/sites/ashoka/2013/08/20/will-obamas-power-africa-initiative-reach-the-rural-poor/. Accessed on 22 November 2014.

International Energy Agency (IEA). (2010). Energy poverty: How to make modern energy access universal. Available at http://www.se4all.org/wp-content/uploads/2013/09/Special_Excerpt_of_WEO_2010.pdf. Accessed on 23 November 2014.

International Energy Agency (IEA). (2011). Energy for all: Financing access for the poor. Special early excerpt of the World Energy Outlook 2011. Available at http://www.worldenergyoutlook.org/resources/energydevelopment/energyforallfinancingaccessforthepoor/. Accessed on 23 November 2014.

———. (2011/2012). World Energy Statistics. Available at http://www.eia.gov/cfapps/ipdbproject/IEDIndex3.cfm?tid=2&pid=2&aid=12. Accessed on 22 November 2014.

International Finance Corporation (IFC). (2012). From gap to opportunity: Business models for scaling up energy access. Available at http://www.ifc.org/wps/wcm/connect/ca9c22004b5d0f098d82cfbbd578891b/EnergyAccessReport.pdf?MOD=AJPERES. Accessed on 22 November 2014.

International Institute for Applied Systems Analysis. (2012). *Global Energy Assessment—Toward a Sustainable Future*. Cambridge, UK and New York, NY: Cambridge University Press.

International Renewable Energy Agency. (2012). International Off-grid Renewable Energy Conference 2012: Key Findings and Recommendations. Available at http://www.irena.org/DocumentDownloads/Publications/IOREC_Key%20Findings%20and%20Recommendations.pdf. Accessed on 23 November 2014.

Jain, S. (2014). The Ghosts of India's power crisis return, *Business Standard*, 9 July 2014.

Jha, S. (2013). UPA rule sees inequality rising in 70% states. *Business Standard*, 6 November 2013. Available at http://wap.business-standard.com/wapnew/storypage.php?id=5amp;autono=113051000984&autono=113110600030. Accessed on 23 November 2014.

Kanase Patil, A.B., Saini, R.P., and Sharma, M.P. (2010). Integrated renewable energy systems for off grid rural electrification of remote area. *Renewable Energy*, 35(6), 1342–1349.

Kapur, A. (2012). *India Becoming: A Portrait of Life in Modern India*. New York: Penguin Books.

Khandker, S.R., Barnes, D.F. and Samad, H.A. (2010). Energy poverty in rural and urban India: Are the energy poor also income poor? The World Bank, Policy Research Working Paper 5463. Available at http://elibrary.worldbank.org/doi/pdf/10.1596/1813-9450-5463. Accessed on 22 November 2014.

IUATLD. (2009). Lung health consequences of exposure to smoke from domestic use of solid fuels. International Union Against Tuberculosis and Lung Diseases, Paris, 2009. Available at: http://www.theunion.org/what-we-do/publications/english/pub_indoor-air-pollution_eng.pdf. Accessed on 22 November 2014.

Khandker, S.R., Samad, H.A., Ali, R., and Barnes, D.F. (2012). Who benefits most from rural electrification? Evidence in India. World Bank, Policy Research Working Paper 6095. Available at http://elibrary.worldbank.org/doi/pdf/10.1596/1813-9450-6095. Accessed on 22 November 2014.

Lee, C.M. et al. (2013). Assessing the climate impacts of cookstove projects: Issues in emissions accounting. Stockholm Environment Institute. Working Paper 2013-01. Sweden.

Lvovsky, K. (2001). Health and Environment, environment strategy paper 1. Washington DC: World Bank.

Martinet, E., Cabraal, A. and Mathur, S. 2001. World Bank/GEF Solar Systems projects: Experiences and lessons learned, 1993–2000, *Renewable and Sustainable Energy Review*, 5, 39–57.

Martin, R. (2013). How long will it take to lift one billion people out of poverty? World Bank Policy Research Working Paper No. 6325. Washington, DC: World Bank.

Milinger, M. and Marlind, T. (2010). Factors influencing the success of decentralized solar power systems in remote villages: A case study in Chhattisgarh, India. http://publications.lib.chalmers.se/records/fulltext/155077.pdf. Accessed on 22 November 2014.

Miller, D. (2009). *Selling Solar*. London: Earthscan.
Mishra, A. and Sarangi, G.K. (2011). Off grid energy development in India: An approach towards sustainability. OASYS South Asia Project Working Paper 12. Available at http://www.academia.edu/7465899/Working_Paper_12_Off-grid_energy_development_in_India_An_approach_towards_sustainability. Accessed on 22 November 2014.
Mitavachan, H., Gokhale, A. and Srinivasan, J. (2011). A case study of 3-MW scale grid-connected solar photovoltaic power plant at Kolar, Karnataka: Performance assessment and recommendations. Available at http://www.dccc.iisc.ernet.in/3MWPV_Plant.pdf. Accessed on 22 November 2014.
Modi, V., McDade, S., Lallement, D., and Saghir, J. (2005). Energy and the millennium development goals. New York: Energy Sector Management Assistance Programme, United Nations Development Programme, UN Millennium Project, and World Bank.
Mohit, P. (2013). KPMG, Annual Conference & Exhibition. Off Grid Solar Summit 'Potential of Solar Off-grid Applications in India', Crowne Plaza Today, New Delhi, 5 July.
Morris, S. et al. (2006). A scheme for efficient subsidisation of kerosene in India. IIM Ahmedabad. Available at http://www.iimahd.ernet.in/publications/data/2006-07-06smorris.pdf. Accessed on 22 November 2014.
Neil, S. and Thomas, P. (2012). India's "energy trilemma": An international perspective. In *Empowering Growth: Perspectives on India's Energy Future*. A report from the Economist Intelligence Unit. Available at http://www.economistinsights.com/sites/default/files/downloads/Empowering_Growth.pdf. Accessed on 22 November 2014.
Ninan, T.N. (2014). *Business Standard*, 12 July.
Pachauri, S., Rao, N.D., Nagai, Y. and Riahi, K. (2012). *Access to Modern Energy: Assessment and Outlook for Developing and Emerging Regions*. Laxenburg, Austria: IIASA.
Palit, D. and Chaurey, A. (2011). Off-grid electrification experience in South Asia: Status and best practices (May). OASYS-South Asia Project: Working Paper Series. Available at http://oasyssouthasia.dmu.ac.uk/docs/oasyssouthasia-wp1-oct2010.pdf. Accessed on 22 November 2014.
Palit, D., Malhotra, R. and Kumar, A. (2011). Sustainable model for financial viability of decentralized biomass gasifier based power projects. *Energy Policy*, 39(9), 4893–4901.
Panagariya, A. (2013). An open letter to Rahul. *Times of India*, 18 November 2013. Available at http://lite.epaper.timesofindia.com/getpage.aspx?pageid=10&pagesize=&edid=TOI-H&edlabel=TOIH&mydateHid=18-11-2013&pubname=Times+of+India+-+Hyderabad&edname=Hyderabad&publabel=TOI. Accessed on 23 November 2014.
Panagariya, A. and Merkins, M. (2013). A comprehensive analysis of poverty in India. World Bank, Policy Research Working Paper 6714, December.
Parikh, J. (2005). The energy poverty and gender nexus in Himachal Pradesh, India: The impact of clean fuel access policy on women's empowerment. Available at http://www.energia.org/fileadmin/files/media/pubs/2005_finrep_parikh.pdf. Accessed on 22 November 2014.
Patil, B. (2010). Modern Energy Access to All in Rural India: An Integrated Implementation Strategy. Discussion paper 2010-08, KSG, Harvard University.
Prahlad, C.K. and Hart, S.L. (2002). The fortune at the bottom of the pyramid, *strategy +business*, issue 26, first quarter. Available at http://www.strategy-business.com/article/11518?pg=all. Accessed on 22 November 2014.
Prayas Energy Group. (2011). Thermal power plants on the anvil: Implications and the need for rationalisation. Discussion paper, Prayas Energy Group, Deccan Gymkhana, India.
———. (2012). Decentralized renewable energy (DRE) Micro-grids in India: A review of the recent literature. Prayas Energy Group, Pune. Available at http://www.prayaspune.

org/peg/publications/item/187-decentralised-renewable-energy-dre-microgrids-in-india.html. Accessed on 22 November 2014.

Rajiv Gandhi Grameen Vidyutikaran Yojana Scheme. (2005). Ministry of Power, Government of India. Available at http://rggvy.gov.in/rggvy/rggvyportal/index.html. Accessed on 22 November 2014.

Ramana, M.V. (2012). *The Power of Promise: Examining Nuclear Energy in India*. New Delhi: Viking Publication.

Reddy, Amulya, K.N. (2009). Energy technologies and policies for rural development. In Thomas B. Johansson and José Goldemberg (eds) *Energy for Sustainable Development: A Policy Agenda*. Available at http://www.undp.org/content/dam/aplaws/publication/en/publications/environment-energy/www-ee-library/sustainable-energy/energy-for-sustainable-development-a-policy-agenda/Energy%20for%20Sustainable%20Development-PolicyAgenda_2002.pdf. Accessed on 22 November 2014.

Report of the High Powered Committee on Financial Position of Oil Companies. (2008). Prime Minister's Office. Available at http://www.indiaenvironmentportal.org.in/files/B%20K%20Chaturvedi%20Report.pdf. Accessed on 22 November 2014.

Shankar, A. (2013). Give Bharat its share of electricity. *Economic Times*, 1 July. Available at http://articles.economictimes.indiatimes.com/2010-07-01/news/27570521_1_electricity-implementation-programme. Accessed on 22 November 2014.

Subramaniam, A. (2013). The economic consequences of Professor Amartya Sen: Redistributive policies via rights and entitlements are ultimately self-defeating. *Business Standard*, 10 July. Available at http://www.business-standard.com/article/opinion/the-economic-consequences-of-professor-amartya-sen-113070901024_1.html. Accessed on 22 November 2014.

Shunglu Committee. (2011). Report of the high level panel on financial position of distribution utilities, Government of India. Available at http://planningcommission.gov.in/reports/genrep/index.php?repts=hlpf.html. Accessed on 22 November 2014.

Singal, S., Varun and Singh, R. (2007). Rural electrification of a remote island by renewable energy sources. *Renewable Energy*, 32(15), 2491–2501.

Smith, K.R. (1998). Indoor air pollution in India: National health impacts and cost effectiveness of intervention. United Nations Development Programme Capacity 21 Project. Mumbai: Indira Gandhi Institute of Development Research.

Sreekumar, N. and Santanu, D. (2011). Rajiv Gandhi Rural Electrification Program: Urgent need for mid-course correction. Discussion paper by Prayas Group. Available at http://www.prayaspune.org/peg/publications/item/162-rajiv-gandhi-rural-electrification-program-urgent-need-for-mid-course-correction.html. Accessed on 24 November 2014.

Subramaniam, A. (2013). The economic consequences of Professor Amartya Sen: Redistributive policies via rights and entitlements are ultimately self-defeating; *Business Standard*, 10 July. Available at http://www.business-standard.com/article/opinion/the-economic-consequences-of-professor-amartya-sen-113070901024_1.html. Accessed on 23 November 2014.

Summary Report Vienna Energy Forum. (2013). Available at http://www.unido.org/fileadmin/user_media/Services/Energy_and_Climate_Change/Renewable_Energy/VEF_2013/summary_report.pdf. Accessed on 22 November 2014.

Sutter, C. (2003). Sustainability check-up for CDM projects: How to assess the sustainability of international projects under Kyoto Protocol. Swiss Agency for Development and Cooperation, ISBN 3-936846-59-6, October.

Telecom Regulatory Authority of India (TRAI). (2011). Recommendations on approach towards green telecommunications. Available at http://www.trai.gov.in/WriteReadData/

Recommendation/Documents/Green_Telecom-12.04.2011.pdf. Accessed on 23 November 2014.

The 18th Electric Power Survey; Central Electricity Authority. (2011).

Tranum, S. (2013). *Powerless: India's Energy Shortage and Its Impact*. New Delhi: SAGE Publications, p. 191.

United Nations. (2002). Johannesburg Plan of Implementation.

———. (2010). The UN Secretary-General's Advisory Group on Energy and Climate Change; Summary Report and Recommendations. New York: United Nations.

———. (2012). The future we want. Available at http://www.uncsd2012.org/content/documents/727The%20Future%20We%20Want%2019%20June%201230pm.pdf. Accessed on 23 November 2014.

Venkataraman, C., Sagar, A.D., Habib, G., Lam, N. and Smith, K.R. (2010). The Indian national initiative for advanced biomass cookstoves: The benefits of clean combustion. *Energy for Sustainable Development*, 14, 63–72.

Von Shrinding, Y., Bruce, N., Smith, K., Ballard-Tremeer, G., Ezzati, M. and Lvovsky, K. (2010). Addressing the impact of household energy and indoor air pollution on the health of the poor—Implications for policy action and intervention measures. Geneva: World Health Organization. Available at http://www.who.int/mediacentre/events/H&SD_Plaq_no9.pdf?ua=1. Accessed on 22 November 2014.

World Bank. (2004). The impact of energy on women's lives in rural India. ESMAP. Washington, DC: World Bank. Available at https://www.esmap.org/sites/esmap.org/files/The%20Impact%20of%20Energy%20on%20Women's%20Lives%20in%20Rural%20India.pdf. Accessed on 23 November 2014.

———. (2010). Empowering Rural India: Expanding Electricity Access by Mobilizing Local Resources. Available at http://siteresources.worldbank.org/INDIAEXTN/Resources/empowering-rural-india-expanding-electricity-access-by-mobilizing-local-resources.pdf. Accessed on 23 November 2014.

———. (2011a). Household Cookstoves, Environment, Health, and Climate Change. Washington, DC: World Bank. Available at http://www.cleancookstoves.org/resources_files/household-cookstoves.pdf. Accessed on 22 November 2014.

———. (2011b). India: Biomass for sustainable development: Lessons for decentralised energy delivery: Village energy security programme. Available at http://www.indiawaterportal.org/sites/indiawaterportal.org/files/India_Biomass_for_Sustainable_Development_World_Bank_2011.pdf. Accessed on 23 November 2014.

———. (2014). More power to India: The challenge of electricity distribution. Directions in development. Washington DC. Available at http://documents.worldbank.org/curated/en/2014/06/19703395/more-power-india-challenge-electricity-distribution. Accessed on 22 November 2014.

Zhang, Y. (2010). Finding out the Killer in the Kitchen: An Analysis of Household Energy Use, indoor Air Pollution, and Health Impacts in India. Saarbrücken, Germany: Lap Lambert Academic Publishing.

Zhang, Y., Barnes, D.F., Sen, M. and Naumoff, K. (2006). Preliminary results: Indoor air pollution and development in India. Working Paper, University of Maryland, College Park, Baltimore.

Index

above poverty line (APL), 59, 60, 62–64
Accelerated Rural Electrification Programme (AREP), 46, 47
ACS. *See* average cost of supply
acute lower respiratory infection (ALRI), 164
acute respiratory infection, 165
Advisory Group on Energy and Climate Change (AGECC), 13
Agriculture Development and Training Society (ADATS), 216
Air Quality Guidelines (AQGs), 168
Appropriate Rural Technology Institute (ARTI), 185
Arashi Hi-Tech Bio-Power Private Limited (AHBPPL), 113
Asian Development Bank (ADB), 143
Association of Power Producers (APP), 39
Atomic Energy Centre (AEC), 30
Automatic Teller Machine (ATM), 234
auxiliary nurse midwives (ANMs), 233
average cost of supply (ACS), 37
average revenue realised (ARR), 37

balance of system (BOS), 117
Base Transceiver Stations (BTSs), 232
below poverty line (BPL), 45, 47, 49, 61

Bihar Renewable Energy Development Agency (BREDA), 131
biogas, 179–81
biogas development, national project, 188–94
biomass, 179
British Petroleum (BP), 126
Bureau of Indian Standards (BIS), 186

carbon market, 216–18
Central Electricity Authority (CEA), 52
Central Electricity Regulatory Commission (CERC), 34, 136
Central Electronics Limited (CEL), 157
Central Public Sector Enterprise (CPSE), 157b
Centre for Science and Environment (CSE), 81, 82, 154
Chhattisgarh Renewable Energy Development Agency (CREDA), 96, 98
chronic obstructive pulmonary disease (COPD), 165, 168
Clean Development Mechanism (CDM), 216–18
Coal India Limited (CIL), 44
community health centres (CHCs), 233
compact fluorescent lamp (CFL), 3, 75, 97

Container Corporation of India (CONCOR), 157
cooking energy
  biogas, 179–80
  biogas development, national project, 188–93
  biomass, 179
  collection and cooking time, 170–71
  forestation, 169–70
  global burden, 174–75
  health impacts, 164–69
  IAP and poverty, 171–72
  improved cookstove initiative, 186–87, 187b–88b
  kerosene, 180
  liquid petroleum gas, 180, 193–95
  national efforts, 180–81
  national programme, 199–202
  National Programme, on Improved Cookstoves, 183–86
  policy implications, 172–74
  problem, 163–64
  solar cooking, 195–95
  status in India, 175, 176f, 177t–78t
Corporate Social Responsibility (CSR), 81, 126, 127, 156–58, 157b
Credit Guarantee Trust for Medium and Small Enterprises (CGTMSEs), 106

Decentralized Distributed Generation (DDG), 35, 49, 94t–95t
Department of Atomic Energy (DAE), 30
detailed project report (DPR), 75, 160
Deutsche Gesellschaft für Internationale Zusammenarbeit (GIZ), 98, 143, 200
Deutsche Gesellschaft für Technische Zusammenarbeit (GTZ), 171
diesel generator (DG), 231
Disability-adjusted Life Year (DALY), 166

distribution companies (Discoms), 22, 36, 38f
domestic funding, 219

Electricity Act 2003, 48
energy access
  agricultural pump sets, 236–39
  background, 1–3
  human development, 4–9, 5f, 7f, 8f
  impact, 239
  international efforts for, 12–18
  programmes and RGGVY, 57
  public places, 235–36
  rural banking, 234
  rural education, 234–35
  rural energy market, 229–30
  rural enterprises, 230–31
  rural health centres, 233–34
  service centres, 235–36
  small commercial markets, 231
  telecommunications, 231–33
The Energy and Resources Institute (TERI), 70, 89
energy consumption, 7f
Energy Development Index (EDI), 6, 7f

Framework Convention on Climate Change (FCCC), 217
Funding. *See* subsidies and funding

generation-based incentive (GBI), 136
giga watt (GW), 20
global scenario, 10–12
Gokak Committee Report, 57–58
Grameena Abhivrudhi Mandali (GAM), 112
greenhouse gas (GHG), 170
gross domestic product (GDP), 4, 25

home lighting system (HLS), 73–75
Human Development Index (HDI), 4, 5f
Husk Power Systems (HPS), 103–06

India electricity status
  coal, 25–27
  CO_2 emission, 21f
  conventional sources, 25
  cost of electricity supply, in rural areas, 35b
  cost of generation, 33–34
  cost of power, 33–40
  cost of supply, 34
  current electricity mix, 22–23, 22f, 23f
  electricity consumption, 21f
  implications, 32–33
  Integrated Energy Policy (IEP), 23
  large hydro, 28–29
  natural gas, 27–28
  nuclear, 29–30
  renewable power installed capacity, 24f
  renewables, 30–32
  tariff issues, 36–37
  technology-wise electricity mix, 24f
  total commercial energy supply (TCES), 20
  utilities, financial situation of, 37–40, 37t
Indian Council of Medical Research, New Delhi (ICMR), 169
Indian Credit Ratings Agency (ICRA), 38
Indian Energy Exchange (IEX), 39
Indian Institute of Technology (IIT), 186, 200
Indian Renewable Energy Development Agency (IREDA), 87, 102
Indian Tobacco Company (ITC), 105
Indo-Norwegian project, 99
indoor air pollution (IAP), 165, 168
Industrial Training Institutes (ITIs), 159
Institute of Minerals & Materials Technology (IMMT), 186

Integrated Energy Policy (IEP), 23, 26, 28, 29
Integrated Research and Action for Development (IRADe), 59
internal rate of return (IRR), 237
international funding, 218
International Monetary Fund (IMF), 213, 243
International Union Against Tuberculosis and Lung Disease (IUATLD), 164, 165, 172

Karnataka Power Transmission Corporation Limited (KPTCL), 35b
Kartnataka Electricity Regulatory Commission (KERC), 35b
kerosene, 180, 207–12
Klynveld Peat Marwick Goerdeler (KPMG), 237

Left Wing Extremism (LWE), 90
light emitting diode (LED), 3, 75, 90
Lighting a Billion Lives (LaBL), 89
liquefied natural gas (LNG), 28
liquid petroleum gas (LPG), 2, 175, 180, 193–95, 205–207

Mahatma Gandhi National Rural Employment Guarantee Act (MGNREGA), 244
Millennium Development Goals (MDGs), 13
Minimum Needs Program (MNP), 45
Ministry of New and Renewable Energy (MNRE), 73
Ministry of Power (MOP), 136, 140

National Action Plan on Climate Change (NAPCC), 30
National Agricultural Bank and Rural Development (NABARD), 47, 82

Index

National Biogas and Manure Management Programme (NBMMP), 181
National Council of Applied Economic Research (NCAER), 225
National Electricity Policy (NEP), 50
National Improved Cookstove Initiative (NCI), 186
National Programme on Improved Cookstoves (NPIC), 181, 183
National Project on Biogas Development (NPBD), 188
National Sample Survey (NSS), 175
National Sample Survey Organisation (NSSO), 175
National Thermal Power Corporation (NTPC), 33, 39
non-communicable diseases (NCDs), 174
non-governmental organisation (NGO), 88, 99, 100, 125–26
non-resident Indian (NRI), 103

Observer Research Foundation (ORF), 138, 159
Omni-grid Micropower Company (OMC), 101, 130
Orb energy, 88–89

PDS. *See* Public Distribution System (PDS)
poverty link basket (PLB), 68
Power Finance Corporation (PFC), 47, 140
Pradhan Mantri Gramoday Yojana (PMGY), 46
primary health centres (PHCs), 233
Promoting Energy Access through Clean Energy (PEACE), 143
Public Distribution System (PDS), 154
public sector undertaking (PSU), 81, 93

Rajiv Gandhi Grameen Vidyutikaran Yojana (RGGVY), 49–52, 53–56, 53t
Rajiv Gandhi Gramin LPG Vitaran Yojana (RGGLVY), 193
Remote Village Electrification (RVE), 75, 80–81
renewable energy
  Avani Initiative of Pine Needle-based Power Plant, 114
  bank-financed solar lighting programme, 81–90, 84t–86t
  biomass technology, 116
  franchisee model, 110–11
  lanterns, 108–9
  localised grid-based supplies, 111–12
  Malavalli Biomass Power Plant, 112
  micro hydro, 116–17
  mini-grid model, 90–107
  Off-grid Renewable Energy Systems, 115
  1.25 MW Biomass Gasification Power Project, 113
  1.2 MW Grid-connected, Biomass Gasification Power Plant, 113–14
  One Megawatt Wood Biomass-based Gasifier Plant, 112–13
  remote village electrification programme, 74–81, 76t–78t
  small wind generators/wind-PV hybrids, 118
  solar photovoltaic technologies, 117–118
  Three Megawatt Solar PV Plant, 114–15
Renewable Energy Development Fund (REDF), 141
Renewable Energy Technology (RET), 35b
Reserve Bank of India (RBI), 141
Rural Electricity Distribution Backbone (REDB), 49

Rural Electricity Supply Technology (REST), 46
rural electrification
  critique of approach, 67–69
  definition, 65–66
  Electricity Act 2003, 48
  franchisee arrangements, 69–71, 70f
  geographical dispersal, 63–64
  of hamlets, 62
  hours of supply, 61–62
  household connectivity, 55–56, 56t
  household electrification, 62–63
  issues in, 60–69
  policies (1974–2005), 45–48
  public facilities, 65
  transformer capacity, 64–65
Rural Electrification Corporation (REC), 43
Rural Infrastructure Development Fund (RIDF), 47

Sagar Rural Energy Development Cooperative (SREDCOP), 102
Saran Renewable Energy (SRE), 106–07
scheduled caste (SC), 45
scheduled tribe (ST), 45
Self Employed Women's Association (SEWA), 185
self-help group (SHG), 197
single point power supply (SPPS), 69b
Small Industries Development Bank of India (SIDBI), 140
solar cooking, 195–96
Solar Electric Light Company (SELCO), 125, 141
solar home system (SHS), 81, 96
solar photovoltaic (SPV), 75
State Electricity Board (SEB), 43, 45
State Electricity Regulatory Commission (SERC), 35b
subsidies and funding
  for biogas, 213–15
  carbon market, 216–18
  cooking energy, 205
  for cookstoves, justification, 212–13
  domestic funding, 219
  financing of, 215–16
  international funding, 218
  kerosene, 207–12
  kerosene conundrum, 219–20, 221t–23t, 224–27
  LPG, difficulties of, 205–207
  World Bank view, 212
Sustainable Energy for All (SE4ALL), 14

Tamil Nadu State Electricity Board (TNEB), 113
technical backup unit (TBU), 183
Technology Informatics Design Endeavour (TIDE), 185
television (TV), 97, 103, 121
TERI. *See* The Energy and Resources Institute (TERI)
tonne oil equivalent (toe), 6
total commercial energy supply (TCES), 20

United Nations Conference on Sustainable Development (UNCSD), 15
United Nations Development Programme (UNDP), 6, 14, 112
United Nations Industrial Development Organisation (UNIDO), 14
United Progressive Alliance (UPA), 49
United States Agency for International Development (US AID), 88
Universal electricity access
  alternative framework, 144–48
  bank-financed programmes, 152–53
  bilateral/multilateral agencies, 143–44

comprehensive mapping, 150–60
corporate social responsibility, 156–58, 157b
costs, 131–37
energy access trust fund, 155–56
entrepreneurial challenges, 125–28
grid interactivity, 137–39
institutional structure, 149–50
organisational challenges, 124–25
regulatory barriers, 131
sociological challenges, 124–25
subsidy support, 131–37
tariff, 131–37
technical challenges, 128–30
universal electricity access, 17f

Village Electrification Infrastructure (VEI), 49
Village Energy Committees (VECs), 91, 92f
Village Energy Security Programme (VESP), 90–91, 99

West Bengal Renewable Energy Development Agency (WBREDA), 102
willingness to pay (WTP), 92, 97, 103
World Health Organization (WHO), 165, 167, 168
World Summit on Sustainable Development (WSSD), 13

About the Authors

P.C. Maithani is currently working as Director in the Ministry of New and Renewable Energy. During his professional career, he has worked on different aspects of renewable energy programmes and policies. Dr Maithani was Visiting Fellow with TERI for a year in 2006 and is also Visiting Faculty at TERI University. He holds doctorate in Physics from HNB Garhwal University and Post Graduate Diploma in Public Policy and Management from Indian Institute of Management, Bangalore. He has contributed many papers and articles and authored a book *Renewable Energy in the Global Context*.

Deepak Gupta served as Secretary to the Ministry of New and Renewable Energy, Government of India, from 1 July 2008 to 30 September 2011. He was instrumental in the launching of India's ambitious national Solar Mission. He belongs to the 1974 batch of the Indian Administrative Service (IAS). A postgraduate in History from St Stephen's College and MPhil in International Relations from Jawaharlal Nehru University, he did a Masters in Public Administration from the Kennedy School, Harvard University, in 1992 as a Mason Fellow. He has worked in different areas in the central and state governments, including a deputation to India Trade Centre, Brussels, in the eighties. He has also served as Adviser with the World Health Organization in Delhi in 2004 for TB. He has extensively worked on rural energy issues and had designed programmes for achieving universal energy assess. He has consulted for ECOWAS (West African group of countries), World Bank and Rural Livelihoods Programme of Government of India for renewable energy and energy access after retirement. He was invited as expert in international conferences on energy access. He also authored two books related to his earlier assignments—*Covering a Billion with DOTS* and *A Documentary Study of Participatory Irrigation Management*.